20-38

2 05/N

DOUBLY TIMELIKE SURFACES

by

John K. Beem

and

Peter Y. Woo

DOUBLY TIMELIKE SURFACES

CONTENTS

DOUBLY TIMELIKE SURFACES*

INTRODUCTION

This paper represents parts of the doctoral dissertations done by the authors at the University of Southern California under Busemann. The authors would like to express their thanks to him for his help and encouragement.

An axiomatic, purely geometric approach has been used by Busemann and Beem to study indefinite metric spaces in a paper "Axioms for Indefinite Metrics". A space satisfying the axioms given in that paper is called a general G-space.

Here we denote the indefinite distance from x to y by xy. The term indefinite distance is slightly misleading since the case where $xy = 0$ implies $x = y$ is included as a general G-space. Thus an indefinite distance is one that is not necessarily definite.

Because of several difficulties, see Busemann and Beem [7], the geodesics are not defined in terms of the distance. Instead, rather mild topological conditions (T) are introduced and then we postulate (A_{1-5}) the existence of a set A of arcs playing the role of sufficiently small arcs of the extremals. These arcs then determine the analogues to the extremals which we call paths. The metric axioms (M_{1-5}) guarantee that for doubly timelike surfaces the paths are in fact extremals.

The conditions (T) and (A_{1-5}) guarantee that a two dimensional space is a topological manifold. Furthermore, convex neighborhoods exist in the two dimensional case.

Received by the editors August 29, 1968.

* This work was supported in part by the National Science Foundation, under contract number GP 7472.

The arcs of A are denoted by Arcs. They are either isometric to a segment of the real axis or else all distances are zero along them. In the latter case they are called null Arcs. A path determined by null Arcs is called a null path.

In the first Chapter we discuss general G-spaces and their geometric covering spaces. A two dimensional general G-space with exactly two null elements (or directions) at each point is called a doubly timelike surface. A doubly timelike surface always has the plane as its topological covering space. An important example is exterior hyperbolic geometry. This space is the two dimensional analogue to both the hyperbolic and the elliptic geometry.

Let A^2 denote the real two dimensional affine space. If the doubly timelike surface R is taken as A^2 and $xy = 2wx = 2wy$ whenever $w = \frac{1}{2}(x + y)$, we call R a Minkowski space. If z denotes the origin in A^2, then the unit circle K in R is $\{x | xz = 1\}$. The light cone C is $\{x | xz = 0\}$. The best known example of a Minkowski plane is the two dimensional Lorentz space also known as the pseudo-euclidean plane.

In all doubly timelike Minkowski planes K has four strictly convex non-compact components and C consists of two affine lines that may be taken as the coordinate axes. Symmetric perpendicularity implies the components of K are differentiable and asymptotic to the lines of C. A Minkowski plane with a pairwise transitive group of motions is shown to be nearly Lorentzian.

In Chapter Three we discuss several short topics. If R is A^2 and the ordinary translations are motions, then R is Minkowskian. In analogy to a theorem in [3], we show a doubly timelike surface with zero curvature is locally Minkowskian.

Surprisingly, curves in doubly timelike surfaces may not have length (not even ∞). However, convex curves can always be assigned length. We define rectifiability for curves and show a convex curve of finite length is rectifiable exactly when it does not contain a null Arc as a subcurve.

An intrinsic theory of differentiability is given in analogue to a theory developed by Busemann in [4]. We make use of this theory in the chapter on Flat Bisectors.

In Chapter Four we define a doubly timelike plane to be a doubly timelike surface homeomorphic to A^2. In a plane the null paths are divided into two families each of which cover the space simply.

A path-unique space is one such that for every pair of distinct points there is exactly one path through both of them. It is shown that each path is either homeomorphic to the real line or a circle. Furthermore, a path unique doubly timelike surface is either homeomorphic to the Möbius strip or a plane. If it is homeomorphic to a plane it is called a straight plane.

Chapter Five deals with perpendiculars, asymptotes, limit circles and parallels in a straight plane. Each straight plane can be represented by a subset S of the (x,y)-plane, whose intersection with any line parallel to either the x or y axis is exactly a null path. Much of the theory is quite analogous to that in G-spaces, but there is one theorem requiring a different proof: A Desarguesian straight plane in which the parallel axiom holds and all circles are concave is Minkowskian.

The last Chapter is on Flat Bisectors. The bisector of the points a_1 and a_2 is defined as $B(a_1,a_2) = \{x \mid xa_1$ and xa_2 both exist and are equal$\}$. Let R be a doubly timelike surface in which locally $B(a_1,a_2)$ contains a certain subset which is a segment whenever $a_1a_2 > 0$. Then we say

R satisfies the Bisector Condition.

If R does satisfy this condition it is locally symmetric and may be reflected (locally) in segments. For any two points $p,q \in R$ there are neighborhoods $U(p)$ and $U(q)$ that are topologically isometric. If R also satisfies a differentiability condition, then it is either locally Minkowskian or locally exterior hyperbolic. This differentiability condition is probably not necessary.

A locally Minkowskian space satisfying the bisector condition is nearly locally Lorentzian. It is locally Lorentzian exactly when (locally) it may be reflected in null Arcs.

In this last Chapter we give the analogue to the Pythagorean Theorem in exterior hyperbolic geometry. This demonstrates one of the advantages of (locally) considering distances to be defined in all directions. Locally a doubly timelike surface consists of two independent timelike spaces in which the strict reverse triangle inequality holds. Most previous investigations of indefinite metrics such as [23] and [5] have not emphasized the study of these two timelike spaces taken together.

DOUBLY TIMELIKE SURFACES

CHAPTER I

GENERAL G-SPACES

1. Timelike Metrics. In this paper the (indefinite) distance from x to y will be denoted by xy. In all cases xy will be defined on a subset Π_0 of the cartesian product $R \times R$ which contains the diagonal Θ. The metric or distance xy will satisfy $xx = 0$, $xy \geq 0$ and (if both xy and yx are defined) $xy = yx$.

Let $x \leq y$ be a partial ordering of $R \times R$ and let Π_0 consist of the pairs (x,y) with $x \leq y$. The distance is timelike if $xy > 0$ for $x < y$ and $xy + yz \leq xz$ for $x < y < z$. This last inequality is called the Time Inequality. Timelike spaces have been studied in some detail by Busemann [5].

The best known example of a timelike space is the Lorentz space where R is the n dimensional real affine space A^n, moreover in suitable affine coordinate with

$$\lambda_n(x) = x_1^2 - \sum_{i=2}^{n} x_i^2 \qquad \text{for} \quad x = (x_1, \cdots, x_n)$$

the partial ordering $x < y$ is given by $x_1 < y_1$ and $\lambda_n(x - y) > 0$. The distance is given by $xy = \lambda_n^{\frac{1}{2}}(x - y)$.

Although A^n has the natural topology $x_\nu \to x$ if $x_\nu^i \to x^i$ for all $i = 1,2,\cdots,n$ it is often useful to have a definite metrization. A euclidean metrization $e(x,y)$ of A^n is associated to A^n if it is invariant under the translations of A^n, or equivalently, if in any system of affine coordinates

$$e(x,y) = \left[\sum_{i,k=1}^{n} g_{ik}(x^i - y^i)(x^k - y^k) \right]^{\frac{1}{2}}, \quad g_{ik} = g_{ki}$$

where

$$\sum_{i,k=1}^{n} g_{ik} x^i x^k$$

is a suitable positive definite quadratic form. The different associated $e(x,y)$ all induce, of course, in A^n the natural topology.

2. General G-Spaces. Busemann and Beem have given a synthetic approach to the study of indefinite metrics in [7]. A space satisfying the axioms given in [7] is called a general G-space. We give in this section the requirements T, A_{1-5} and M_{1-5} necessary for R to be a general G-space.

The space R must satisfy (at least) the following topological requirements.

T. R is a locally compact, connected Hausdorff space with a countable base.

It is known, see [18, p. 74] and [21] that R can be metrized such that it becomes finitely compact, i.e. so that bounded infinite sets have accumulation points. Although this metric is not unique, it proves useful to select one such distance $\delta(x,y)$ once and for all as an auxiliary device. The open ball $\{x| \ \delta(x,p) < \rho\}$ is denoted $S(p,\rho)$.

A system A of Jordan arcs is assumed given in R satisfying A_{1-5} (a point is not an arc).

A_1. If $\alpha \in A$ then every subarc of α also lies in A.

A_2. Each arc α lies in the relative interior of some $\alpha' \in A$.

A_3. If α and α' lie in A and have an endpoint and further point in common, then either $\alpha \subset \alpha'$ or $\alpha' \subset \alpha$.

The axiom A_3 implies at most one arc in A exists joining two distinct points x and y. We denote the symmetric set of all pairs of endpoints of arcs in A by Π and by $\alpha(x,y)$ the arc in A from x to y when $(x,y) \in \Pi$.

A_4. If $(x_n,y_n) \in \Pi$, $x_n \to x$, $y_n \to y$ and $(x,y) \in \Pi_0 = \Pi \cup \Theta$. Then $\alpha(x_n,y_n) \to \alpha(x,y)$ if $x \neq y$ and $\alpha(x_n,y_n) \to x$ if $x = y$.

Here the limit is Hausdorff's closed limit.

A_5. Π_0 is an open subset of $R \times R$.

For brevity we denote the arcs in A by Arcs.

It follows from A_5 that each point p has a neighborhood U such that any two distinct points in U can be connected by an Arc. In terms of the auxiliary metric $\delta(x,y)$ there is a $\rho > 0$ such that $S(p,\rho) \times S(p,\rho) \subset \Pi_0$. Let $\delta(p)$ be the least upper bound of these ρ. Then, see [3, p. 33], $S(p,\delta(p)) \times S(p,\delta(p)) \subset \Pi_0$ and either $\delta(p) = \infty$, i.e. $\Pi_0 = R \times R$ or

$$|\delta(p) - \delta(q)| \leq \delta(p,q).$$

A topology for the Arcs is introduced by defining a neighborhood U_α of $\alpha(x,y)$ to consist of all Arcs $\alpha(x',y')$ for which x' and y' are, respectively in certain neighborhoods U_x of x and U_y of y. It follows from A_5 that $U_x \times U_y \subset \Pi_0$ for sufficiently small U_x and U_y. We deduce by A_4 that $\alpha_n \to \alpha$, if and only if, each U_α contains all but a finite number of α_n. Thus the limit induced by the U_α for the Arcs coincides with the Hausdorff closed limit.

The Arcs are divided into classes called paths, where α and α' belong to the same path, if and only if, a finite number of Arcs $\alpha = \alpha_1, \alpha_2, \cdots, \alpha_n = \alpha'$ exist such that $\alpha_i \cap \alpha_{i+1}$ for $i = 1,2,\cdots,n-1$ is an Arc.

A line element λ at a point q is a maximal set of Arcs containing q in their interiors such that any two Arcs in λ have an Arc in common.

If P is a path through q and one Arc of λ lies on P, then they all do. We denote this by $\lambda \subset P$ and call λ a line element of P at α. The multiplicity of P at q is the cardinal number of distinct line elements of P at q. A point of multiplicity one is a simple point of P, all others are called multiple points. If all points of P are simple, P is called simple. If $x,y \in P$ implies $(x,y) \in \Pi_0$ then P is called a line.

A representation of a path P is a map $x(t)$ of a connected open set I_P of the real axis into R with the following two properties:

a) For each Arc α in P numbers $t_1 < t_2$ exist such that $x(t)|[t_1,t_2]$ is a topological map of α on $[t_1,t_2]$.

b) Given $t_0 \in I_P$, then $t_1 < t_0 < t_2$ in I_P exist such that $x(t)|[t_1,t_2]$ is a topological map of $[t_1,t_2]$ on an Arc in P.

It can easily be shown that each path possesses a representation.

A number of properties of spaces satisfying T and A_{1-5} are listed in [7]. Here we are concerned with two dimensional general G-spaces and mention only the next two results.

(I.1) If $\dim R = 1$ or 2, then R is a topological manifold.

A non-empty set C in R is convex if $C \times C \subset \Pi_0$ and $\alpha(x,y) \subset C$ for distinct $x,y \in C$.

(I.2) If $\dim R = 2$ each point p has a convex neighborhood.

If $(x,y) \in \Pi$ and $z \in \alpha(x,y) - x - y$, we write (xzy).

The metric axioms M_{1-5} that a general G-space must satisfy are now given.

M_1: A continuous function xy is defined on Π_0 which satisfies $xx = 0$, $xy = yx$ and is bounded on any subset of Π_0 with compact closure in $R \times R$.

M_2: If $x_0 y_0 = 0$ for one pair of distinct points of an Arc α in A, then $xy = 0$ for all $x,y \in \alpha$.

M_3: If $xy > 0$ for one pair (hence all pairs) of distinct points on $\alpha \in A$, then xy defines on α a (definite) metric with which α is isometric to an interval of the real axis.

If $xy = 0$ for all x,y on α we call α a null Arc, otherwise a segment. A representation $z(s)$ of a segment α is a map of an interval I of the real axis onto α such that $z(s_1)z(s_2) = |s_1 - s_2|$ for all $s_1, s_2 \in I$.

Let q be a point and α a segment with $(q,x) \in \Pi_0$ for all $x \in \alpha$. The pair (q,α) is called

<u>Spacelike,</u> if $qx > 0$ for $x \in \alpha$ and $|qx - qy| < xy$ for distinct $x,y \in \alpha$.

<u>Neutral,</u> if $|qy - qx| = xy$ for $x,y \in \alpha$ unless u with (xuy) and $qu = 0$ exists.

<u>Timelike,</u> if qx has at least one zero in the interior of α and $|qx - qy| > xy$ or $|qx' - qy'| > x'y'$ whenever $qu = 0$ and x,y,u,x',y' lie in this order on α.

It is easily shown that if (q,α) is neutral, then qx has at most one zero on α. If (q,α) is timelike, qx has at most two zeros on α.

M_4: Let $U(p)$ be a neighborhood of p for which $\bar{U}(p) \times \bar{U}(p) \subset \Pi_0$. Then a neighborhood $V(p) \subset U(p)$ exists such that for any point $q \in V(p)$ and segment α which intersects $V(p)$, lies in $\bar{U}(p)$ and has its endpoints on $\bar{U}(p) - U(p)$, the pair (q,α) is spacelike, timelike or neutral.

M_5: If p is an interior point of the segment α, then neighborhoods U_α of α and $W(p)$ of p exist such that for $q \in W(p)$ and $\alpha' \in U_\alpha$ the pair (q,α') is either spacelike or is neutral with one zero of qx in the interior of α' or is timelike with two zeros of qx in the interior of α'.

If one Arc of a path or a line element is a null Arc they all are. In that case we call the path a null path or the line element a null element.

A path all (or one) of whose Arcs are segments is called a <u>geodesic</u>. Among the representations of a geodesic G some have the form $z(s)$, $s \in I_G$ where for each $\alpha(x,y) \subset G$ a number u exists such that $z(s)|[u,u+xy]$ represents α. The parameter s is then arclength.

The term "representation of a geodesic" will henceforth be reserved for this form. If $I_G = (-\infty,\infty)$ for one representation of G, then this holds for all representations.

A general G-space is <u>geodesically complete</u> if for each geodesic G the set I_G is the entire real axis.

If R is a general G-space and S an open, connected subset of R, then we may make S into a general G-space by the following:

For each $p \in S$ let $W(p)$ be a neighborhood of p such that $W(p) \subset S$, $W(p) \times W(p) \subset \Pi_0$. Let $V(p) \subset W(p)$ be a neighborhood of p with $x,y \in V(p)$ implying $\alpha(x,y) \subset W(p)$. Define $\Pi_0(S) = \bigcup_{p \in S} V(p) \times V(p)$. Then if (x,y)

belongs to $\Pi_0(S)$ let the Arc in S from x to y be the original $\alpha(x,y)$ and let the distance in S from x to y be the original xy.

Consider such an open connected set S in R. We call S path convex if all paths in S (considered as space) are homeomorphic to the real line and if any two points $x,y \in S$ can be joined by a unique path in S.

A distance preserving map of one general G-space into another is not necessarily one to one or continuous. Consequently, we introduce the concept of a topological isometry. Let M_i be a subset of the topological space R_i and let a function $\delta_i(x,y)$ be defined on $M_i \times M_i$ with $\delta_i(x,x) = 0$ and $\delta_i(x,y) = \delta_i(y,x) \geq 0$. The map Φ of M_1 onto M_2 is a topological isometry if it is topological and

$$\delta_2(\Phi(x),\Phi(y)) = \delta_1(x,y) \quad \text{for all } x,y \in M_1.$$

3. Two Dimensional Spaces. The two dimensional general G-spaces fall into four categories, see [7, p. 241].

(I.3) Theorem. Let $\dim R = 2$. Then there is either no null element, or exactly one at each point, or there are exactly two at each point, or all line elements are null elements.

The interesting cases occur when there are no null elements and when there are exactly two null elements at each point. A geodesically complete general G-space with no null elements is nearly a G-space, see [7, p. 239]. A two dimensional general G-space with exactly two null elements at each point is called a *doubly timelike surface*.

Let C be an open convex subset of a doubly timelike surface. Then the null lines in C considered as space, can be divided into two disjoint families F_M and F_N such that the lines in each family cover C simply. If $x \in C$ we denote the elements of F_M and F_N through x by $M(x)$ and $N(x)$.

The families F_M and F_N form sets homeomorphic to a line and hence each have two natural orderings or orientations. Let F_M^+ and F_N^+ be orientations of F_M and F_N.

(I.4) Let C be an open convex subset of a doubly timelike surface. Using the above notation define two partial orderings $x <_i y$ on C as follows:

$x <_1 y$ means $M(x) < M(y)$ and $N(x) < N(y)$; and $x <_2 y$ means $M(x) < M(y)$ and $N(y) < N(x)$. For each $x \leq_i y$, the given distance xy defines in C a timelike space, i.e. $xy + yz \leq xz$ if $x <_i y <_i z$ and equality holds only if (xyz).

Note that any two points x,y in C satisfy exactly one of the relations

$$xy = 0, \quad x <_1 y, \quad y <_1 x, \quad x <_2 y, \quad y <_2 x.$$

For reference we state another useful result found in [7].

(I.5) Let (q,α) be timelike with two zeros u_1, u_2 of qx on α. If distinct u_1, u_2, x, y lie in this order on α then $qy > qx + xy$ and $qx > qu_2 + xu_2$.

(I.6) Let L_1 and L_2 be two null lines in the convex set C with $L_1 \cap L_2 = u$. If $q \in L_1 - u$ and u,x,y lie in this order on L_2, then $qy > qx$.

<u>Proof</u>: It is only necessary to show $qy > qx$ for all y sufficiently close

to x with (uxy). Let L(q,x) denote the line through q and x, then for y close to x we have (y,L(q,x)) a timelike pair, see M_4. Choose $x_1 \varepsilon \alpha(q,x) - q - x$. Then

$$yq > yx_1 + x_1q \geq yx + xx_1 + x_1q = yx + xq.$$

4. Covering Spaces. Let R be a general G-space. Then R is connected, locally connected and locally simply connected. Thus R satisfies the conditions for which the theorems of covering spaces as described by Pontryagin [17] and Seifert and Threlfall [20] are true, in particular, the following: For each class of conjugate subgroups of the fundamental group F(R), there exists a unique covering space $(\bar{R},\Omega)$ over R such that the covering map Ω induces an isomorphism of $F(\bar{R})$ onto some subgroup of the conjugate class.

Definition: The map $\phi: R_1 \rightarrow R_2$ is called a local isometry if for each point p in R_1 there is a neighborhood U(p) such that $\phi|U(p)$ is a topological isometry.

A homeomorphism that is a local isometry is called an isometry.

Let (R,Ω) be a covering space of R. We shall construct $\Pi_0(\bar{R})$, and Arc system, and a metric on $\bar{R}$ so that it becomes a general G-space with Ω a local isometry. We then call $(\bar{R},\Omega)$ a geometric covering space of R. Without mentioning it every time, a subset or a point in $\bar{R}$ is denoted by a letter with an upper bar, and the corresponding letter without the bar always means the image in R under Ω.

Let $\bar{x}$ be any point of $\bar{R}$. Let $\bar{U}(\bar{x})$ be a neighborhood of $\bar{x}$ such that $\Omega|\bar{U}(\bar{x})$ is a homeomorphism and $U(x) \times U(x) \subset \Pi_0$. Let V(x) be a neighborhood of x such that $p,q \in V(x)$ implies $\alpha(p,q) \subset U(x)$. Then for $p,q \in V(x)$ define $\alpha(\bar{p},\bar{q}) = [\Omega|\bar{U}(\bar{x})]^{-1}\alpha(p,q)$ and define $\Pi_0(\bar{R}) = \bigcup_{\bar{x}\in\bar{R}} \bar{V}(\bar{x}) \times \bar{V}(\bar{x})$ where $\bar{V}(\bar{x}) = [\Omega|\bar{U}(\bar{x})]^{-1}V(x)$. Also define $\overline{pq} = pq$. Then $(\bar{R},\Omega)$ is a geometric covering space of R.

A trivial example is when $\bar{R} = R$ and Ω is the identity mapping. Then $\Pi_0(\bar{R}) \subset \Pi_0$ and $\Pi_0(\bar{R})$ is not necessarily equal to Π_0, but if P is a path in R it is also a path in $\bar{R}$.

(I.7) Theorem. For each conjugate class of subgroups of the fundamental group $F(R)$ of a general G-space R, there is up to isometry only one geometric covering space $(\bar{R},\Omega)$ associated with the conjugate class topologically, $\bar{R}$ being metrized so that Ω is a local isometry.

Proof: Topologically, $(\bar{R},\Omega)$ exists and can be metrized as above. Suppose there were two geometric covering spaces $(\bar{R}_1,\Omega_1)$ and $(\bar{R}_2,\Omega_2)$ of R such that for each $i = 1,2$ the map Ω_i induces an isomorphism from $F(\bar{R}_i)$ into one of the subgroups in the conjugate class. Then from [17, p. 356] we know there exists a homeomorphism ϕ from $\bar{R}_1$ onto $\bar{R}_2$ with $\Omega_1 = \Omega_2\phi$. Since Ω_1 and Ω_2 preserve distances locally, it follows ϕ preserves distances locally and hence is an isometry.

It follows that there exists one and only one geometric universal covering space $(\bar{R},\Omega)$ of a general G-space. The next three results follow as in [3].

(I.8) A locally isometric map of a space $\bar{R}$ onto another space R is an isometry if $F(\bar{R})$ is not isomorphic to any proper subgroup of $F(R)$.

(I.9) If $F(R)$ is not isomorphic to any proper subgroup of itself, then any locally isometric map of R onto itself is a motion (defined below).

(I.10) There is a motion Φ of the universal covering space $\bar{R}$ over any motion ϕ of R, i.e., for each $x \in R$ and x^* in $\Omega^{-1}(x)$, we have $\phi(x) = \Omega\Phi(x^*)$.

Topologically a two dimensional manifold can have only the two sphere or the plane as universal covering space. If a doubly timelike surface had the sphere as universal covering space, Theorem (IV.1) would imply the existence of a continuous vector field on the two sphere. Since this is impossible we have the next result.

(I.11) Theorem. A doubly timelike surface has the plane as universal covering space.

5. Motions. A motion of a metric space R is an isometry of R onto itself. The condition that the map be onto is important because a metric space may have isometries into proper subsets.

Definition: A motion of the general G-space R is an isometry of R onto itself.

Thus, a motion of R is a homeomorphism Φ of R onto R such that for each $p \in R$ there is a neighborhood V(p) on which Φ is a topological isometry. Any Arc in V(p) is mapped onto an Arc in R. However, it may happen that $(x,y) \in \Pi_0$ yet $(\Phi(x), \Phi(y)) \notin \Pi_0$. Any path P of R is mapped onto a path P'.

From the definition it is clear that if Φ is a motion, then so is Φ^{-1}.

It follows that the motions of R form a group.

6. Example. Exterior Hyperbolic Geometry. We give now an important example of a general G-space which will be referred to later.

Let x^0, x^1, x^2 be homogeneous coordinates for the real two dimensional projective space P^2. Define

$$\varepsilon_i = \begin{cases} -1 & \text{if } i = 0 \\ +1 & \text{if } i = 1,2 \end{cases}$$

The space R is then the subset of P^2 given by $\varepsilon_i (x^i)^2 > 0$. If x is a point of R having a representation x^0, x^1, x^2 we normalize by choosing a representation $\underline{x}^0, \underline{x}^1, \underline{x}^2$ of x such that

$$\varepsilon_i (\underline{x}^i)^2 = 1. \quad \text{(Summation over i)}$$

For two points x and y in R either $|\varepsilon_i \underline{x}^i \underline{y}^i| > 1$ or $|\varepsilon_i \underline{x}^i \underline{y}^i| \le 1$. In the first case set

$$xy = k_1^{-1} \cosh^{-1} |\varepsilon_i \underline{x}^i \underline{y}^i| \quad \text{for } k_1 > 0.$$

In the second case set

$$xy = k_2^{-1} \cos^{-1} |\varepsilon_i \underline{x}^i \underline{y}^i| \quad \text{for } k_2 > 0.$$

Let x and y be distinct points of R. If $|\varepsilon_i \underline{x}^i \underline{y}^i| \ge 1$ the projective line $L(x,y)$ through x and y has one segment from x to y that lies entirely in R. This segment will be an Arc and be denoted $\alpha(x,y)$. If $0 < |\varepsilon_i \underline{x}^i \underline{y}^i| < 1$, then $L(x,y)$ contains one segment from x to y such that any p on this segment has $xp \le xy$. This segment will be an Arc and be denoted $\alpha(x,y)$.

The above distance restricted to the indicated Π_0, together with the above Arc system, makes R a general G-space, see [23, p. 60]. We denote this space by $EH(k_1,k_2)$ and shall call it an exterior hyperbolic geometry.

In [23] the two cases of xy treated above are considered separately. By treating the two cases of xy together to get $EH(k_1,k_2)$, we are able to formulate an analogue to the Pythagorean Theorem, see (VI. 33).

Another important example is an analogue to the Poincare model of hyperbolic geometry discussed by Busemann and Beem in [7]. For reference the space is denoted by R^h and the system of Arcs by A^h.

R^h is the half plane $x_2 > 0$ of the (x_1,x_2)-plane. The paths are the parts in $x_2 > 0$ of the following curves

a.) $x_2 = \pm x_1 + \beta$

b.) $-(x_1 - \beta)^2 + x_2^2 = \gamma, \quad \gamma > 0$

c.) $(x_1 - \beta)^2 - x_2^2 = \gamma, \quad \gamma > 0$

d.) $x_1 = \beta$

In case b.) a path is a branch of a hyperbola, in c.) we have one-half of each of the two branches of a hyperbola. These two pieces count, of course, as different paths. The Arcs are the subarcs of the paths, so that the paths are lines.

Consider $ds_h = |dx_1^2 - dx_2^2|\,x_2^{-1}$ in R^h. The curves a.) of R^h are evidently the null paths. The extremals for the variational problem with $(dx_1^2 - dx_2^2)x_2^{-1}$ as integrand are the curves in b.) and those of $(dx_2^2 - dx_1^2)x_2^{-1}$ are the curves in c.). The distance xy is defined for $(x,y) \in \Pi_0$ as the length of the Arc $\alpha(x,y)$ in terms of ds_h.

The system of paths, A^h and ds_h are invariant under the quasi-hyperbolic group [3, p. 371]. Its elements are therefore motions of R^h.

It can be shown, see [22], that each point of R^h has a neighborhood topologically isometric to an open set of EH(1,1). For a discussion of the universal covering space of R^h see Woo, [22].

The doubly timelike analogue to the euclidean plane is the pseudo-euclidean plane. Let R be A^2 and define the Arcs to be the affine segments. For $x = (x_1, x_2)$ and $y = (y_1, y_2)$ set

$$(x,y) = x_1 y_1 - x_2 y_2 .$$

With this indefinite inner product R is known as the pseudo euclidean plane and as the two dimensional Lorentz space.

The space R is a doubly timelike surface if xy is defined by

$$xy = |(x-y, x-y)|^{1/2}$$

Consider the partial ordering $x < y$ given by $x_1 < y_1$ and $(x-y, x-y) > 0$. If $x < y < w$ then $xy + yw \leq xw$. Equality holds only for y on the affine segment from x to w. This reverse triangle inequality expresses the clock paradox of special relativity. The distance xy is interpreted as the proper time (eigen-time) it takes to go from event x to event y traveling at constant velocity.

DOUBLY TIMELIKE SURFACES

CHAPTER II

MINKOWSKI SPACES

1. Preliminary Results. In Finsler spaces, and more generally in G-spaces which are continuously differentiable, see [4], the local geometry is Minkowskian in the same sense the local geometry of a Riemannian space is euclidean. Thus, Minkowski metrics may be considered as the prototypes of (differentiable) definite metrics, and hence have been studied in detail.

A similar situation prevails for non-definite spaces with a non-definite analogue to Minkowski spaces. A two dimensional definite Minkowski space has a unit circle consisting of a compact convex curve. If R is a Minkowski space which is a doubly timelike surface, the unit circle of R has four convex non-compact components.

The definition of Minkowski spaces for general G-spaces is given in [2]. We give here the definition for the case of doubly timelike surfaces.

Definition: The doubly timelike surface R is called a Minkowski space if R is A^2 (with its usual topology), A consists of affine segments and $(x,y) \in \Pi_0$ with $w = \frac{1}{2}(x + y)$ implies $wx = wy = \frac{1}{2}xy$.

The set Π_0 need not equal $R \times R$ but the Arcs and metric may always be extended in a natural fashion so that $\Pi_0 = R \times R$, see [2]. Therefore, we assume $\Pi_0 = R \times R$ when R is a Minkowski space.

Given distinct $x,y \in R$ the Arc from x to y is the affine segment from x to y. The path determined by x and y is the affine line $L(x,y)$ through x and y.

In the rest of this chapter R is assumed to be a doubly timelike surface which is a Minkowski space.

(II.1) For a given affine line L in R there is a number $\phi(L)$ such that $xy = \phi(L)e(x,y)$ for all $x,y \in L$.

Proof: If α is an Arc lying on L, then either α is a null Arc or a segment. If α is null, then L is a null path and $\phi(L) = 0$.

If α is a segment, there is a number $\phi(\alpha)$ with $xy = \phi(\alpha)e(x,y)$ for all $x,y \in \alpha$, compare [3, p. 94]. Given another segment $\beta \subset L$, there is a similar number $\phi(\beta)$. Let α' be a segment containing α and β, then $\phi(\alpha) = \phi(\beta) = \phi(\alpha')$. Setting $\phi(L) = \phi(\alpha)$ yields the desired result.

(II.2) $\phi(L)$ is continuous.

Proof: Let $\{L_\nu\}$ be a sequence of lines converging to L. Choose $x_0,y_0 \in L$ with $x_0 \neq y_0$. There are sequences $\{x_\nu\} \to x_0$ and $\{y_\nu\} \to y_0$. Since $x_\nu y_\nu \to x_0 y_0$ and $e(x_\nu,y_\nu) \to e(x_0,y_0)$ we have $\phi(L_\nu) \to \phi(L)$.

(II.3) Let L_1 and L_2 be parallel lines in R. Then $\phi(L_1) = 0$ implies $\phi(L_2) = 0$.

Proof: Choose $p \in L_1$ and let L_3 be the null line through p with $L_1 \cap L_3 = p$. Let $L_3 \cap L_2 = q$. If L_2 is not a null line, there is a null line L_4 with $L_4 \cap L_3 = q$. Set $r = L_4 \cap L_1$. Then the (convex) triangle with vertices p,q and r would have all sides null contradicting [7, p. 244].

(II.4) If L_1 is a line parallel to L_2, then $\phi(L_1) = \phi(L_2)$.

Proof: Assume $\phi(L_1) > 0$. Choose $a,b \in L_1$ with $ab > 0$ and b following a in some orientation L_1^+ of L_1. Orient L_2 positive in the same direction as L_1^+. Let N_1 and N_2 be the null lines through a. Set $a' = N_1 \cap L_2$ and $u' = N_2 \cap L_2$. Assume a' follows u' on L_2^+. Let $b' = M_1 \cap L_2$ for M_1 the parallel to N_1 through b.

Let $\{p_\nu\}$ traverse L_2^+ in the positive sense with $p_1 = b'$. Then by (I.4) we have $ap_\nu > a'p_\nu$. Hence,

$$1 > \frac{a'p_\nu}{ap_\nu} = \frac{a'b'}{ar_\nu} \quad \text{where} \quad r_\nu = L(b,b') \cap L(a,p_\nu).$$

Thus, $ar_\nu \to ab$ yields

$$1 \geq \frac{a'b'}{ab} = \frac{\phi(L_2)}{\phi(L_1)} .$$

The same reasoning yields $1 \geq \frac{\phi(L_1)}{\phi(L_2)}$ so that $\phi(L_1) = \phi(L_2)$.

(II.5) The affine translations are motions of R.

<u>Proof</u>: Let T be a translation. If $L = L(x,y)$ then $T(L)$ is another line L' parallel to L. Therefore, $xy = \phi(L)e(x,y) = \phi(L')e(T(x),T(y)) = T(x)T(y)$. Since T is a topological isometry taking affine lines to affine lines, T is a motion.

Let z denote the origin. If N_1 and N_2 are the two null lines through z the set $R - N_1 \cup N_2$ has four components S_1, S_2, S_3 and S_4. We may assume $-S_1 = S_3$ and $S_2 = -S_4$. The unit circle $K = \{x \mid xz = 1\}$ then has four components K_1, K_2, K_3 and K_4. Exactly one component lies in each S_i. Without loss of generality we may assume N_1 and N_2 are the coordinate axes and $K_i \subset S_i$ for $i = 1,2,3,4$.

(II.6) The components of K are strictly convex.

<u>Proof</u>: Let $a,b \in K_1$ with $a \neq b$. Set $u = a + b$ and $w = \frac{1}{2}(a + b)$. Then in some partial ordering $<_i$ we have $z <_i a <_i u$ and $z <_i w <_i u$. Also $az = bz = au = 1$. By (I.4) we have $za + au < zu = 2zw$. Thus $2 < 2zw$ and

$\alpha(z,w) \cap K_i$ is a point in the relative interior of $\alpha(z,w)$. It follows that the set $K_1^0 = \{x | \ zx > 1\}$ is strictly convex with boundary K_1.

Take N_1 and N_2 as the x and y axis respectively. Let K_1 be given by the topological map $p(t)$ for $-\infty < t < \infty$. Then $e(z,p(t)) \to \infty$ as $|t| \to \infty$. If $T(t)$ is a supporting line of K_1 at $p(t)$, then $T(t) \to T$ as $t \to +\infty$ and $T(t) \to T'$ as $t \to -\infty$. The sets T and T' are affine lines each parallel to one of the coordinate axes. If T is parallel to the x-axis, either T is the x-axis or $T \cap S_1 \neq \phi$.

(II.7) Let K_1 be a set in the first quadrant of the (x,y)-plane. There is a Minkowski space R with the coordinate axes as null lines and K_1 as one component of the unit circle if, and only if,

1.) K_1 is given by $y = f(x)$ for $x > \gamma$ with $\gamma \geq 0$.

2.) $f(x) \to \infty$ as $x \to \gamma^+$.

3.) $f(x)$ is strictly convex and positive.

2. Isometries of Minkowski Spaces. Let R_1 and R_2 be doubly timelike Minkowski spaces defined on A^2. Let R_i have metric $m_i(x,y)$ for $i = 1,2$. A map T of R_1 onto R_2 such that $m_2(T(x),T(y)) = m_1(x,y)$ for all $x,y \in R_1$ can readily be shown to be a topological isometry. Because of (I.4) it follows that T must map affine segments to affine segments. Consequently, T must be an affinity.

Let K and K^* be the unit circles according to $m_1(x,y)$ and $m_2(x,y)$ respectively. If there is a centro affinity L taking K onto K^*, then clearly $m_1(x,y) = m_2(L(x),L(y))$ for all x,y. Conversely, a centro-affinity L that is an isometry of R_1 onto R_2 maps K onto K^*.

We have shown the following result.

(II.8) Let R_1 and R_2 be Minkowski spaces defined on A^2 with unit circles K and K^*. Then R_1 and R_2 are topologically isometric if, and only if, K and K^* are affinely equivalent.

3. Perpendicularity. Perpendicularity is first defined for doubly timelike surfaces. For Minkowski spaces, the definition is a natural generalization of the definition in pseudo-euclidean spaces.

Definition: Let (q,α) be a timelike pair. The point f is a foot of q on α if qx has two zeros u_1, u_2 on α and qx has a local maximum at $f \in \alpha(u_1, u_2)$.

In Minkowski spaces if L is not a null line then q will have exactly one foot f on L.

Definition: Let α and β be intersecting segments with $\alpha \cap \beta = f$, for f in the relative interior of β. Then α is called perpendicular to β, written $\alpha \dashv \beta$, if there is a neighborhood $U(f)$ such that all $y \in \alpha \cap U(f) - f$ have f as a foot on β.

Null Arcs are always considered perpendicular to themselves.

For Minkowski spaces, two lines L_1 and L_2 with $L_1 \cap L_2 = f$ have $L_1 \dashv L_2$ if each $y \in L_1 - f$ has f as a foot on L_2.

If $L_1 \dashv L_2$ and $L_2 \dashv L_1$, then we write $L_1 \perp L_2$. Perpendicularity is symmetric if $L_1 \dashv L_2$ implies $L_1 \perp L_2$.

(II.9) Let $L_1 \dashv L_2$ and let L_{i+2} be parallel to L_i for $i = 1,2$, then $L_3 \dashv L_4$.

<u>Proof</u>: Translation invariance of the metric xy implies the result.

Take the null lines through f as the coordinate axes. If the ith quadrant is dilated by λ the component K_i of K in this quadrant is mapped to a corresponding component $\lambda K_i = \{\lambda x \mid x \in K_i\}$ of the circle of radius λ about z. A supporting line L of K_i at p is mapped to a parallel λL that is a supporting line of λK_i at λp.

(II.10) <u>Theorem</u>. If perpendicularity is symmetric in R, the components of the circles are differentiable.

<u>Proof</u>: Since any circle of radius r about z can be obtained from the unit circle K be a dilation of factor r^{-1}, it suffices to show the components of K are differentiable.

Assume K_1 was not differentiable at p. Let L_1 and L_2 be two supporting lines of K_1 at p. Then $L_0 = L(z,p)$ must be perpendicular to both L_1 and L_2. Let L_{i+2} for $i = 1,2$ be the parallel to L_i through z. Set $L_3 \cap K_2 = q_1$ and $L_4 \cap K_2 = q_2$ where K_2 is the component of K in the second quadrant. Let L_5 and L_6 be the parallels to L_0 through q_1 and q_2 respectively. Since $L_5 \perp L_3$ and $L_6 \perp L_4$ the lines L_5 and L_6 must be supporting lines of K_2. But the strict convexity of K_2 then implies $q_1 = q_2$ and hence $L_1 = L_2$.

Consider two doubly timelike Minkowski metrics $m_1(x,y)$ and $m_2(x,y)$ defined on A^2. The two metrics lead in general to two different spaces R_1 and R_2. If one component of the unit sphere in R_1 is the same as some component of the unit sphere in R_2, the two spaces have the same families of null lines.

(II.11) <u>Theorem</u>. Let R_1 have symmetric perpendicularity and a unit circle

K with components K_1, K_2, K_3 and K_4 where $K_1 = -K_3$. If R_2 has K_1 as a component of its unit circle, then perpendicularity is symmetric in R_2 if, and only if, for some $\lambda > 0$ the unit circle in R_2 has $K_1, \lambda K_2, K_3$ and λK_4 as components.

<u>Proof</u>: If the unit circle in R_2 has components $K_1, \lambda K_2, K_3$ and λK_4, then clearly perpendicularity is symmetric in R_2. Let K^* denote the unit circle in R_2.

Let perpendicularity be symmetric in R_2. Assume $p \in K^* - K_1 \cup K_3$. The line $L(z,p)$ has a parallel L which is tangent to K_1 at some point q. Consider L_p the parallel to $L(z,q)$ through p. Since $L(z,q) \perp L$ implies $L_p \perp L(z,p)$, the line L_p is tangent to the (differentiable) component of K^* containing p. It follows that given one $p \in K^* - K_1 \cup K_3$ the set K^* is determined. Consequently, K^* must have components $K_1, \lambda K_2, K_3, \lambda K_4$.

This theorem means that given one component, say K_1 of the unit circle in a doubly timelike Minkowski space with symmetric perpendicularity, the unit sphere K is determined up to a constant multiple of K_2 and K_4. A construction using the polar reciprocal of K_1 can be used to find K to within this multiple of K_2 and K_4, see [2, p. 66].

Let a doubly timelike Minkowski space be given with unit circle having components K_1, K_2, K_3 and K_4 where $K_1 = -K_3$. We can find sets K_2^* and K_4^* such that the set with components K_1, K_2^*, K_3 and K_4^* is a doubly timelike Minkowski space with symmetric perpendicularity if, and only if, the sets T and T' defined preceding (II.7) are the null lines through z, see [2, p. 68].

4. Pairwise Transitive Mobility. An ordinary metric space with distance $d(x,y)$ is said to have a transitive group of motions if, whenever two points p,q are given, there exists a motion of the space taking p to q. It is said to have a pairwise transitive group of motions if given two pairs of points p_1,p_2 and q_1,q_2 with $d(p_1,p_2) = d(q_1,q_2)$, there is a motion taking $p_i \to q_i$ for $i = 1,2$. The ordinary or definite Minkowski spaces having pairwise transitive groups of motions can be determined by finding the ones such that for p and q with $d(z,p) = d(z,q) = 1$ there is a motion holding z fixed taking p to q.

The situation is more complicated in indefinite spaces because two distinct points can have zero distance and in higher dimensions K can have components that are not homeomorphic, see [2].

Definition: An indefinite Minkowski space is said to have a pairwise transitive group of motions if given p and q on the same component of K, there always exists a motion holding z fixed, taking p to q.

The reflection in z is always a motion. Therefore, if R has a pairwise transitive group of motions and p,q are points on opposed components of K, there is a motion taking p to q. To determine the doubly timelike Minkowski planes with pairwise transitive groups of motions we first state a lemma found in [5].

(II.12) In the (x,y)-plane let $y = g(x)$ be a strictly decreasing and differentiable curve on (γ,∞) where $\gamma \geq 0$ and $g(x) \to \infty$ for $x \to \gamma^+$. Let the tangent at $q \in (x,g(x))$ intersect the axes at q_x and q_y. If $\Theta \geq |q - q_x| : |q - q_y| \geq \Theta_0 > 0$, then $\gamma = 0$ and $\lim_{x\to\infty} g(x) = 0$. If $\Theta = \Theta_0$ then $g(x) = \lambda x^{-\Theta}$ for $0 < x < \infty$.

This lemma leads to the following two results.

(II.13) Theorem. Let R be a doubly timelike Minkowski plane with a pairwise transitive group of motions. Then affine coordinates x,y exist with the coordinate axes as null lines and with K given by $y = x^{-\Theta}$ in the first quadrant and $y = \lambda|x|^{-\Theta}$ in the second.

Proof: Each component of K is differentiable at some point since it is convex. Since any point on a component can be moved to any other point by a motion, the whole component must be differentiable. Take coordinates such that each component K_i of K lies in the ith quadrant.

If q and $q' \in K_1$, then there is a centro affinity mapping q to q' that is a motion of R. It leaves the ratio $|q - q_y|:|q - q_x|$ of (II.12) fixed, and hence in proper coordinates $y = x^{-\Theta}$ gives K_1. By the same arguments $y = \lambda|x|^{-\psi}$ gives K_2. It remains to show $\Theta = \psi$.

Since $y = x^{-\Theta}$ gives K_1 and since R possesses a pairwise transitive group of motions $K_2 \cup K_4$ must be invariant under the centro affinities $T(x,y) = (cx, c^{-\Theta}y)$ for $c > 0$. But this only holds if $\Theta = \psi$.

It is clear that λ need not equal one since $y = \lambda|x|^{-\Theta}$ is invariant under centro affinities of the above form. There will exist a motion of R taking K_1 onto K_2 exactly when $\lambda = 1$. This implies the following theorem.

(II.14) Theorem. Let R be a doubly timelike Minkowski plane. For every set of pairs of points, p_1,p_2 and q_1,q_2 with $p_1p_2 = q_1q_2 \neq 0$ there will exist a motion taking $p_i \to q_i$ for $i = 1,2$ exactly when in suitable affine coordinates K is given by $|y| = |x|^{-\Theta}$ where $\Theta > 0$.

CHAPTER III

DOUBLY TIMELIKE SURFACES

1. Null Boxes. If C is an open convex set in the doubly timelike surface R, let the null lines in C (considered as space) be divided into two families F_M and F_N. If $x \in C$ let $M(x)$ and $N(x)$ denote the elements of F_M and F_N respectively that contain x. For $p \in C$ and a_1, a_2 points of C sufficiently close to p with $a_1 a_2 > 0$ the intersections $M(a_1) \cap N(a_2)$ and $M(a_2) \cap N(a_1)$ will exist. When they do exist, set $b_1 = M(a_1) \cap N(a_2)$ and $b_2 = M(a_2) \cap N(a_1)$.

The open set bounded by $\alpha(a_1,b_1)$, $\alpha(b_1,a_2)$, $\alpha(a_2,b_2)$ and $\alpha(b_2,a_1)$ is convex and will be denoted $Q(a_1,a_2)$. Clearly, we have $Q(a_1,a_2) = Q(b_1,b_2)$. We call $Q(a_1,a_2)$ a null box since it is bounded by four null Arcs. A null box as well as its closure is always convex.

For $q \in Q(a_1,a_2)$ the intersections $q_1 = M(q) \cap \alpha(a_1,a_2)$ and $q_2 = N(q) \cap \alpha(a_1,a_2)$ will each consist of a single point. Let $f(x) = a_1 x$ for $x \in \alpha(a_1,a_2)$ and consider the map $T(q) = (f(q_1), f(q_2))$ for $q \in Q(a_1,a_2)$. This map has the image D: $0 < x < a_1 a_2$ and $0 < y < a_1 a_2$ in the (x,y)-plane. The sets $N(q) \cap Q(a_1,a_2)$ and $M(q) \cap Q(a_1,a_2)$ for fixed q go to line segments parallel to the y and x axis respectively. The segments in $Q(a_1,a_2)$ are mapped by T to either strictly increasing or strictly decreasing curves in D.

If two doubly timelike surfaces R_1 and R_2 are given, choose convex sets C_i in R_i for $i = 1,2$. In C_i choose a_1^i, a_2^i such that $Q(a_1^i, a_2^i)$ exists and such that $a_1^1 a_2^1 = a_2^2 a_1^2$. Then the maps T_1: $Q(a_1^1, a_2^1) \to D$ and T_2: $Q(a_1^2, a_2^2) \to D$ defined as above have the same image D:

$$0 < x < a_1^1 a_2^1, \qquad 0 < y < a_1^1 a_2^1.$$

The set D can now be made into a doubly timelike surface. For $r_1 = (x_1, y_1)$ and $r_2 = (x_2, y_2)$ in D let $r_1 r_2 = 0$ if $x_1 = x_2$ or $y_1 = y_2$. If either $x_1 < x_2$ and $y_1 < y_2$ or $x_2 < x_1$ and $y_2 < y_1$, we set $r_1 r_2 = T_1^{-1}(r_1)\, T_1^{-1}(r_2)$. Let $r_1 r_2 = T_2^{-1}(r_1)\, T_2^{-1}(r_2)$ in all other cases. The Arcs of D are then externals in terms of the metric $r_1 r_2$.

The above illustrates that locally a doubly timelike surface consists of two timelike spaces which have the same null lines, but which are otherwise independent.

Next we prove two simple but important lemmas. Let C be an open convex set in a doubly timelike surface. Then $<_1$ and $<_2$ denote the partial orderings defined on C.

(III.1) Let $x(t)$ and $y(t)$ for $0 \le t \le 1$ be curves in C with $x(t)y(t) > 0$ for all t. If $x(0) <_i y(0)$, then $x(t) <_i y(t)$ for all t.

<u>Proof</u>: If $p, q \in C$ and $p <_i q$ there are neighborhoods $U(p)$ and $U(q)$ such that $p' <_i q'$ for all $p' \in U(p)$ and $q' \in U(q)$. Hence, if $x(t_0) < y(t_0)$ there is a neighborhood $W(t_0) \subset [0,1]$ with $x(t) <_i y(t)$ for $t \in W(t_0)$.

Assume the assertion is false. Let τ be the first t such that $x(t) \not<_i y(t)$. Since $x(\tau)y(\tau) > 0$, it follows $x(\tau) <_j y(\tau)$ for $i \neq j$. This is impossible since then there would be a $W(\tau) \subset [0,1]$ with $x(t) <_j y(t)$ for $t \in W(\tau)$.

If $a, b, c \in C$ with $ab > 0$, $bc > 0$ and $ca > 0$ set $f_a(x) = ax$ for $x \in \alpha(b,c)$; $f_b(x) = bx$ for $x \in \alpha(a,c)$ and $f_c(x) = cx$ for $x \in \alpha(a,b)$. The sum of the number of zeros of f_a, f_b and f_c can equal at most two by [7, p. 244].

(III.2) The sum of the number of zeros of f_a, f_b and f_c is exactly two.

Proof: Let F_M, F_N, $M(x)$ and $N(x)$ be as above, assume without loss of generality f_a has no zeros. For $x \in \alpha(a,b)$ the intersection of $M(x)$ and $\alpha(a,c)$ is either empty or a point y_1. The intersection of $N(x)$ and $\alpha(a,c)$ is either empty or y_2. For all $x \in \alpha(a,b)$ and sufficiently close to a, both y_1 and y_2 will exist. When $x \neq a$ and both y_i exist they are distinct. Furthermore, each map $x \to y_i$ (when y_i exists) is one to one and continuous. There are three significantly different cases to consider.

1.) For all $x \in \alpha(a,b)$ the point c never lies in any $M(x)$ or $N(x)$. Then f_c has no zeros, but $M(b)$ and $N(b)$ both intersect $\alpha(a,c)$ so f_b has two zeros.

2.) There exists $r \in \alpha(a,b)$ with $c \in M(r)$, yet c is never in any $N(x)$. Then f_c has one zero, namely r, and $N(b) \cap \alpha(a,c)$ is the one zero of f_b.

3.) There are two distinct points $r,s \in \alpha(a,c)$ with $c \in M(r) \cap N(s)$. Then f_c has two zeros and f_b none.

2. Translations. A G-space with a transitive abelian group of motions is locally Minkowskian. In general G-spaces this is not always true. Consider the (x,y)-plane with the distance from p to q defined as the absolute value of the difference in their x coordinates. If p and q lie on a translate of the curve $xy = -1$ with $x < 0$, let $\alpha(p,q)$ be the arc from p to q along this translate. In all other cases let $\alpha(p,q)$ be the affine segment from p to q. This yields a general G-space with $\Pi_0 = R \times R$ with the metric and Arcs both invariant under the affine translations. How-

ever, in the case of a doubly timelike surface we have:

(III.3) Let A^2 be given a doubly timelike metric $m(p,q)$. If the affine translations are motions, the space is Minkowskian.

Proof: First we will show that the null Arcs lie on affine segments. Let $e(p,q)$ be an associated euclidean metrization of A^2. Assume α_1 is in $U(z) = \{q \mid e(z,q) < \varepsilon\}$ with $\varepsilon > 0$ and $\bar{U}(z) \times \bar{U}(z) \subset \Pi_0$. Let $p \in \alpha_1$ and $u \in \alpha(z,p)$, then set $u' = p - u$. If u is not on the affine line $L(z,p)$, then u and u' lie on opposite sides of $L(z,p)$. The following holds because of translation invariance $m(z,u') = m(u,p) = 0$ if α_1 is null.

If $\alpha(z,2^{-1}p)$ is extended past $2^{-1}p$ until it first meets the boundary of $U(z)$, at some point it must intersect one of the $u \in \alpha(z,p)$ or else one of the u'. This implies $m(z,2^{-1}p) = 0$ and that $\alpha(z,2^{-1}p)$ is a null Arc. Using the same argument on $\alpha(z,2^{-1}p)$ and translation invariance, one obtains $m(z,v) = 0$ for all $v \in L(z,p)$ with $(z,v) \in \Pi_0$. Since each point has exactly two null elements, it follows that all null Arcs lie on affine segments.

Let L_1 and L_2 be the two affine lines through z with $m(z,v) = 0$ if $v \in L_1 \cup L_2$ and $(z,v) \in \Pi_0$. The set $U(z)$ is divided into four sectors by L_1 and L_2. For $p \in U(z) - (L_1 \cup L_2)$ all $u \in \alpha(z,p) - z - p$ and $u' = p - u$ must lie in the same component of $U(z) - (L_1 \cup L_2)$. Let $u_0 \in \alpha(z,p)$ with $m(z,u_0) = m(u_0,p) = \frac{1}{2} m(z,p)$. Then $m(z,u_0') = m(u_0',p)$ and by (I.4) $u_0 = u_0'$. Therefore, $u_0 = \frac{1}{2} p$. This implies $\alpha(z,p)$ lies on the affine segment from z to p, and that the space is Minkowskian.

3. Zero Curvature. Let R be a general G-space with $(a,b) \in \Pi$ and $ab > 0$. The midpoint of a and b which we denote by m_{ab} or $m[a,b]$ is

then the unique point of $\alpha(a,b)$ such that $am_{ab} = 2^{-1}ab$. For (a_ν, b_ν), and $(a,b) \varepsilon \Pi$ with $a_\nu b_\nu > 0$, $a_\nu \to a$ and $b_\nu \to b$ we have $m[a_\nu, b_\nu] \to m[a,b]$ if $ab > 0$. When $ab = 0$ the sequence $m[a_\nu, b_\nu]$ may not even converge.

Definition: The general G-space R is said to have zero curvature, if for each $p \varepsilon R$, there is a neighborhood $U_1(p)$ such that for $q,r,s \varepsilon U_1(p)$ with $qr > 0$ and $qs > 0$, the relation $2m_{qr}m_{qs} = rs$ holds.

In the rest of this section R will always be assumed to be a doubly timelike surface with zero curvature.

Let $U_1(p)$ be as in the above definition, and choose $U_0(p)$ to be some $Q(a_1, a_2)$ with $\bar{U}_0(p) \times \bar{U}_0(p) \subset \Pi_0$ and $U_0(p) \subset U_1(p)$. Let Δabc denote a triangle in $U_0(p)$ with $ab > 0$, $ac > 0$, and $bc > 0$. The functions $f_a(x)$, $f_b(x)$ and $f_c(x)$ are then as in (III.2). We recall that $U_0(p)$ has two partial orderings $<_1$ and $<_2$, see (I.4).

(III.4) If either f_c or f_b have no zeros $b <_i c$ implies $m_{ab} <_i m_{ac}$.

Proof: Assume f_c has no zeros. Let $x(t)$ for $t \varepsilon [0,ab]$ be the representation of $\alpha(b,a)$ with $x(0) = b$. For all t we have $x(t)c > 0$ so that $m_{x(t)c}$ is always defined. When $t \to 0^+$, then $\alpha(m_{x(t)b}, m_{x(t)c}) \to \alpha(b, m_{bc})$. For sufficiently small t, then $m_{x(t)b} <_i m_{x(t)c}$. Since $m_{x(t)b}m_{x(t)c} = 2^{-1}bc$ for all $t > 0$ it follows from (III.1) that $m_{ab} <_i m_{ac}$.

We wish to show that in Δabc, the relation $b <_i c$ always implies $m_{ab} <_i m_{ac}$. The only case not covered in (III.4) occurs when f_b and f_c each have one zero. We treat this case using the following lemma.

(III.5) Let Δabc be as above and let both f_b and f_c have one zero.

Then if $d \varepsilon \alpha(a,b)$ and $cd > 0$ the point m_{cd} lies on $\alpha(m_{ac}, m_{cb})$.

Proof: The point m_{cd} is defined since $cd > 0$. If $a <_j b$ then (III.4) implies $m_{ac} <_j m_{cd} <_j m_{cb}$. Since R has zero curvature $2m_{ac}m_{cd} + 2m_{cd}m_{cb}$ equals $ad + db$ as well as ba. Hence $m_{cd} \varepsilon \alpha(m_{ac}, m_{cb})$.

(III.6) When Δabc is as above and both f_b and f_c have one zero, $b <_i c$ implies $m_{ab} <_i m_{ac}$.

Proof: Let $x(t)$ be the representation of $\alpha(a,b)$ for $t \varepsilon [0,ab]$ with $x(0) = b$. Set $y(t) = \alpha(x(t),c) \cap \alpha(m_{ac}, m_{bc})$. Then $y(t)$ is defined for $t \varepsilon [0,ab]$ and has image $\alpha(m_{ac}, m_{bc})$. If $x(t)c > 0$, then $y(t) = m_{x(t)c}$ and $m_{x(t)b}y(t) = 2^{-1}bc$. Since $x(t)c = 0$ for only one value of t, we conclude $m_{x(t)b}y(t) = 2^{-1}bc$ for all t. By (III.1) we have $m_{ab} <_i m_{ac}$.

It now follows that if Δabc is given and $d \varepsilon \alpha(b,c)$ with $ad > 0$ then $m_{ad} \varepsilon \alpha(m_{ab}, m_{ac})$ no matter how the zeros of f_a, f_b, and f_c are placed.

(III.7) Let $w,x,y,z \varepsilon U_0(p)$ with $wx > 0$, $xy > 0$, $yz > 0$, $zw > 0$ and $w <_i z$, $x <_i y$. Then $m_{wx} <_i m_{zy}$. Furthermore, $2m_{wx}m_{zy} \geq wz + xy$.

Proof: There exist points $u \varepsilon \alpha(w,z)$ and $v \varepsilon \alpha(x,y)$ such that $wv > 0$, $uv > 0$, $zv > 0$, $xu > 0$ and $yu > 0$. Then

$$m_{wx} <_i m_{xu} <_i m_{uv} <_i m_{vz} <_i m_{zy} \quad \text{and}$$

$$2m_{wx}m_{xu} + 2m_{xu}m_{uv} + 2m_{uv}m_{vz} + 2m_{vz}m_{zy} = wz + xy.$$

Consequently, $2m_{wx}m_{zy} \geq wz + xy$.

This implies the following result.

(III.8) Let $x(t)$ and $y(t)$ be representations of segments and let $\alpha(x(t),y(\Theta t))$ be a segment in $U_0(p)$ for all $\Theta_1 < t \leq \Theta_2$. If either $x(\Theta_1) = y(\Theta\Theta_1)$ or $\alpha(x(\Theta_1),y(\Theta\Theta_1))$ is a segment in $U_0(p)$, the function $\phi(t) = x(t)y(\Theta t)$ is concave on $[\Theta_1,\Theta_2]$.

If $\alpha(a,b)$ and $\alpha(c,d)$ are segments, the linear mapping $x \to x'$ of $\alpha(a,b)$ onto $\alpha(c,d)$ is given by $ax{:}ab = cx'{:}cd$.

(III.9) Let Δabc be as before so that $ab > 0$, $ac > 0$, and $bc > 0$. Then the linear mapping of $\alpha(a,b)$ onto $\alpha(a,c)$ has $x <_i x'$ for $b <_i c$ and $x \in \alpha(a,b) - a$. Furthermore, $xx' = (ax)(bc)(ab)^{-1}$.

<u>Proof</u>: For $t \in [0,ab]$ consider the representation $x(t)$ of $\alpha(a,b)$ with $x(0) = a$. Let $y(t)$ for $t \in [0,ac]$ be the representation of $\alpha(a,c)$ with $y(0) = a$. The linear mapping $x \to x'$ then takes $x(t)$ to $y(\Theta t)$ for $\Theta = (ac)(ab)^{-1}$. The function $\phi(t) = x(t)y(\Theta t)$ for $0 \leq t \leq ab$ is then linear since it is concave and $\phi(0) = 0$, $\phi(2^{-1}ab) = 2^{-1}bc$, and $\phi(ab) = bc$. This establishes both parts of the lemma.

(III.10) <u>Theorem</u>. Let R have zero curvature and Δabc be as above. Then Δabc is isometric to a triangle in a Minkowski space.

<u>Proof</u>: Let x and x' be as above. For $z \in \alpha(x,x')$ set $ax = \xi_z$ and $zx = \eta_z$. Then $\eta_{x'}$ is a linear function of ξ_x. If $u \in \Delta abc$ let $\bar{u} \in \alpha(a,b)$ and $\bar{u}' \in \alpha(a,c)$ be points such that $\xi_u = \xi_{\bar{u}} = \xi_{\bar{u}'}$.

If u and v are points with $\xi_u \neq \xi_v$ and $(\bar{u} \times \bar{v})$ then

$$* \quad \eta_{x'} = \frac{\xi_v - \xi_x}{\xi_v - \xi_u}\eta_{\bar{u}'} + \frac{\xi_x - \xi_u}{\xi_v - \xi_u}\eta_{\bar{v}'}.$$

When $uv > 0$ and w is the point of $\alpha(u,v)$ with $wu{:}uv = x\bar{u}{:}\bar{u}\bar{v} =$

$x'\bar{u}':\bar{u}'\bar{v}'$ we have

$$x'w \geq \frac{\xi_v - \xi_x}{\xi_v - \xi_u}(\eta_{\bar{u}'} - \eta_u) + \frac{\xi_x - \xi_u}{\xi_v - \xi_u}(\eta_{\bar{v}'} - \eta_v)$$

and

$$xw \geq \frac{\xi_v - \xi_x}{\xi_v - \xi_u}\eta_u + \frac{\xi_x - \xi_u}{\xi_v - \xi_u}\eta_v \quad \text{by (III.8)}.$$

Adding these inequalities and using * one obtains

$$xw + wx' \geq xx'.$$

However for u and v sufficiently close

$$x <_i w <_i x'$$

and

$$xw + wx' \leq xx'.$$

Therefore, $xw + wx' = xx'$ and $w \in \alpha(x,x')$. This implies that $uw:uv = (\xi_w - \xi_u):(\xi_v - \xi_u)$.

Interpreting (ξ_z, η_z) as a point of the cartesian (ξ,η)-plane Δabc becomes

$$0 \leq \xi \leq ab = \xi_b \quad \text{and} \quad 0 \leq \eta \leq \frac{\xi}{\xi_b}\eta_b.$$

If Δabc is remetrized by the euclidean distance

$$e(u,v) = [(\xi_u - \xi_v)^2 + (\eta_u - \eta_v)^2]^{\frac{1}{2}}$$

then $$e(u,v) = |\eta_u - \eta_v| \quad \text{if} \quad \xi_u = \xi_v.$$

Furthermore $$\frac{uw}{uv} = \frac{\xi_w - \xi_u}{\xi_v - \xi_u} = \frac{e(u,w)}{e(u,v)} \quad \text{whenever} \quad uv > 0, \quad (uzv),\ \xi_u \neq \xi_v$$

and u and v are sufficiently close together.

It follows readily that segments α lie on cartesian lines. Null Arcs must then lie on cartesian lines by A_3. The above also implies that for every segment α in Δabc a factor $\lambda > 0$ exists such that $uv = \lambda e(u,v)$ for $u,v \in \alpha$. If α is a null Arc, we set $\lambda = 0$ and $uv = \lambda e(u,v)$ still holds. Since we may choose α as the entire intersection of a cartesian line with Δabc the factor λ is the same for segments on the same line. It remains to be shown that for two Arcs α and α_1 on distinct parallel lines, the corresponding factors λ and λ_1 are equal. For reasons of continuity, we may assume α and α_1 are actually segments and lie in the interior of Δabc. Let $\alpha = \alpha(u,v)$ and $\alpha_1 = \alpha(u_1,v_1)$ have equal (and not opposite) directions. To show $\lambda_1 = \lambda$ choose w with (uu_1w) so close to u_1 that $\alpha(w,v)$ intersects $\alpha(u_1,v_1)$ in a point v_2. We may assume also that $uu_1 > 0$, $vw > 0$, then $e(w,v_2): e(w,v)$ equals $e(w,u_1): e(w,u)$ and using (III.9)

$$\lambda_1 e(u_1,v_2) = u_1v_2 = \frac{wu_1}{wu} uv = \lambda \frac{e(w,u_1)}{e(w,u)} e(u,v) = \lambda e(u_1,v_2).$$

4. Curves and Arclength. A curve in a general G-space R is a continuous map $w(t)$ of $[a,b]$ into R. For a partition of $[a,b]$ $\Delta: a = t_0 < t_i < \cdots < t_n = b$ with $(w(t_{i-1}),w(t_i)) \in \Pi_0$ for $i = 1,2,\cdots,n$. Define

$$L(w,\Delta) = \sum_{i=1}^{n} w(t_{i-1})w(t_i).$$

Let $||\Delta|| = \max(t_i - t_{i-1})$, then whenever $||\Delta||$ is sufficiently small $\alpha(w(t_{i-1}),w(t_i))$ will exist for all i. In definite metric spaces, if Δ_ν is a sequence of partitions of $[a,b]$ with $||\Delta_\nu|| \to 0$, then $L(x,\Delta_\nu)$ converges to the length of $w(t)$ which may be ∞. In timelike spaces, see [5, p. 9], $||\Delta_\nu|| \to 0$ again implies $L(w,\Delta_\nu)$ converges to the length of $w(t)$. However, in general G-spaces $L(w,\Delta_\nu)$ may not converge.

Definition: The curve $w(t)$ for $a \leq t \leq b$ has length $L(w)$ if $\lim L(w, \Delta_\nu) = L(w)$ for each sequence of partitions of $[a,b]$ with $||\Delta_\nu|| \to 0$.

If $L(w,\Delta_\nu)$ does not converge when $||\Delta_\nu|| \to 0$ we say $w(t)$ does not have length. An example of a curve $w(t)$ without length is given in [2].

In a definite metric space a curve is called rectifiable if its length is finite. In general G-spaces we wish to have rectifiability also imply that there are no non-degenerate subarcs of zero length, compare [5]. We let $L_{t_1}^{t_2}(w)$ denote the length of $w(t)$ for $t_1 \leq t \leq t_2$.

Definition: A curve $w(t)$ for $a \leq t \leq b$ that has length is said to be rectifiable provided:

1.) $L_{t_1}^{t_2}(w) > 0$ if $w(t_1) \neq w(t_2)$

2.) $L(w) < \infty$.

On a rectifiable curve we can introduce arclength as parameter:

$$s(t) = L_a^t(w), \quad s(a) = 0$$

The function $s(t)$ is continuous, see [2]. Furthermore, $s(t)$ is non-decreasing with $s(t_2) > s(t_1)$ for $t_1 < t_2$ and $w(t_1) \neq w(t_2)$.

The standard representation of a rectifiable curve is defined as follows:

$y(s)$ represents the point $w(t)$ for which $L_a^t(w) = s$. The point $w(t)$ is uniquely determined by s, but in general t is not.

The length of rectifiable curves is, in general, neither upper nor lower semi-continuous. This can be seen by the following sequence of curves defined in the (x,y)-plane.

$$w_n(t) = \begin{cases} (t, t-\frac{k}{2n}) & \text{for } k = 0,2,\cdots,2n \text{ and } \frac{k}{2n} \le t \le \frac{k+1}{2n} \\ (t, \frac{k+1}{2n} - t) & \text{for } k = 1,3,\cdots,2n-1 \text{ and } \frac{k}{2n} \le t \le \frac{k+1}{2n} \end{cases}$$

Then if $w(t) = (t,0)$ for $0 \le t \le 1$, we have $\lim w_n(t) = w(t)$ for all t. If we use $ds = |dxdy|^{1/2}$, then the length of $w_n(t)$ is always one, yet the length of $w(t)$ is zero. If we use $ds = |dx^2 - dy^2|^{1/2}$ the length of $w_n(t)$ is zero, yet the length of $w(t)$ is one.

Let now R be a doubly timelike surface and C be an open convex subset of R with partial orderings $<_1$ and $<_2$. A <u>local T-curve</u> in C is a continuous map $w(t)$ of $[a,b]$ into C such that for fixed i either

1.) $w(t_1) <_i w(t_2)$ for all $t_1 < t_2$ in $[a,b]$

or 2.) $w(t_2) <_i w(t_1)$ for all $t_1 < t_2$ in $[a,b]$.

Busemann has defined and studied local T-curves in locally timelike spaces, see [5, p. 8]. We state here some of the results he has obtained.

(III.11) A local T-curve $w(t)$ in C always has a length $L(w)$. Furthermore, $L(w)$ is the infimum of $L(w,\Delta)$ over all partitions Δ of $[a,b]$.

(III.12) If $w_\nu(t)$ for $a \le t \le b$ are local T-curves in C and $w_\nu(t) \to w(t)$ for each $t \in [a,b]$ where $w(t)$ is a local T-curve in C, then we have $\limsup L(w_\nu) \le L(w)$.

5. <u>Convex Curves</u>. By (I.2) every point p of a two dimensional space R satisfying T and A_{1-5} has a convex neighborhood D. The set D may be chosen homeomorphic to E^2 with $\bar{D} \times \bar{D} \subset \Pi_0$. Then each pair of distinct points $x,y \in D$ determine a unique arc $\alpha_{xy} \subset \bar{D}$ such that $\alpha(x,y) \subset \alpha_{xy}$ and

the endpoints of α_{xy} lie on $\bar{D} - D$. The Arc α_{xy} less its endpoints will be called a line in D and denoted P(x,y). If $\Pi_0 = R \times R$ take D to be the whole space and for $x \neq y$ let P(x,y) denote the unique path containing x and y.

A convex curve in D is a component of the boundary of a proper convex subset of D. A convex curve is always the complete boundary of a convex subset of D. An arc (not necessarily an Arc) is convex if it lies on a convex curve.

A supporting line of a curve C is a line that contains at least one point of C, but does not separate any two points of C. The existence of a supporting line at a point p of a convex curve C is deduced as follows: Orient C, and let x approach p on C from the right. The line P(p,x) revolves (because of convexity) monotonically about p, but for C the boundary of the convex set C^0 the line P(p,x) never contains an interior point y of C^0 with (xpy). Hence P(p,x) tends to a limit P_r which is called the "right tangent" of C at p and is a supporting line of C.

If P_k is a supporting line of C at p_k and $p_k \to p$ from the right, then $\lim P_k = P_r$. This follows from the fact that any line which can be approached by a subsequence of $\{P_k\}$ is a supporting line of C at p.

If the right and left tangents of C at p coincide, then no other supporting lines at p exist and we call C "differentiable" at p, and the unique supporting line a "tangent".

The next result is found in [3].

(III.13) The set of points at which a convex curve C in D is not differentiable is at most countable.

Consider now the situation when R not only satisfies T and A_{1-5}

but is a doubly timelike surface. Orient the convex curve C and let v be to the right of u on C. Let $A(u,v)$ then represent the arc of C connecting u to v such that if $x \in A(u,v) - u - v$, then u comes before x which comes before v in the orientation of C. It may happen that $A(u,v)$ consists of one or more null Arcs, in which case the length of $A(u,v)$ denoted $L(A(u,v))$ is zero. However, if P_r is the right tangent to C at u and P_r is not a null line, then there is a neighborhood $U(u)$ such that all $x \in C \cap U(u)$ and to the right of u satisfy $xu > 0$ and $u <_i x$ for fixed i.

(III.14) Let R be a doubly timelike surface and C be a convex curve in D as above. Let the right tangent P_r to C at u not be a null line. If v is to the right of u on C then $L(A(u,v)) > 0$.

<u>Proof</u>: Since length is additive, it suffices to show the result for v close to u. Assume then that $u <_i x$ for all $x \in A(u,v) - u$. Let P be a supporting line to C at v and let v be so close to u that $P \cap P_r$ is a point w of D with $u <_i w <_i v$.

If $x_0 = u$, $x_1, x_2, \cdots, x_n = v$ follow each other from left to right on $A(u,v)$ denote by y_k the intersection of $P(x_{k-1}, x_k)$ with $\alpha(v,w)$. Then $w <_i y_1 <_i \cdots <_i y_{n-1} <_i v$ lie on $\alpha(w,v)$ and

$$uw + wy_1 \le uy_1 = x_0x_1 + x_1y_1$$

$$x_1y_1 + y_1y_2 \le x_1y_2 = x_1x_2 + x_2y_2$$

$$\cdots$$

$$x_{n-1}y_{n-1} + y_{n-1}v \le x_{n-1}v = x_{n-1}x_n.$$

Adding these relations we find

$$uw + wv \le x_0x_1 + \cdots + x_{n-1}x_n.$$

Therefore,

$$0 < uw + wv \le L(A(u,v)).$$

This lemma implies that a convex curve C is rectifiable unless it contains a subarc that is a null Arc.

Let C: $x(t)$ for $a \le t \le b$ be a convex local T-curve. Assume $x(t_1) <_1 x(t_2)$ for all $t_1 < t_2$. Let C': $x'(t)$ for $a \le t \le b$ a second convex local T-curve with $x'(a) = x(a)$, $x'(b) = x(b)$, $x'(t_1) <_1 x'(t_2)$ for $t_1 < t_2$. If C' is contained in the set bounded by $\alpha(x(a),x(b))$ and C, then $L(C') > L(C)$.

(III.15) Let C be a convex curve in D with endpoints p and q. Let C_ν: $x_\nu(t)$ for $a \le t \le b$ be a sequence of convex local T-curves with $x_\nu(t_1) <_1 x_\nu(t_2)$ for all $t_1 < t_2$. Let $x_\nu(a) \to p$ and $x_\nu(b) \to q$. If each C_ν is in the set bounded by C and $\alpha(p,q)$ then, $L(C_\nu) \to L(C)$ if $C_\nu \to C$.

6. Differentiability. Busemann [4] has developed an intrinsic theory of differentiability in G-spaces. To apply exactly the same definitions to indefinite spaces that he used is unsatisfactory since, for example, R^h would not then be differentiable.

Here we consider an approach possible only in two dimensional spaces with exactly two null elements at each point. For $p \in R$ let $U_0(p)$ be a null box with $\bar{U}_0(p) \times \bar{U}_0(p) \subset \Pi_0$. If $xx' > 0$ and $x,x' \in U_0(p)$ the null box $Q(x,x')$ is contained in $U_0(p)$.

If $q \in U_0(p)$ and α is a segment with endpoints on the boundary of $U_0(p)$ through q then for any point a sufficiently close to q there are unique points $x, x' \in \alpha$ with $q = m_{xx'}$ and $a \in \bar{Q}(x,x') - Q(x,x')$. For $a \neq q$ let $\bar{\alpha}(q,a)$ be the minimal Arc such that $\alpha(q,a) \subset \bar{\alpha}(q,a)$ and $\bar{\alpha}(q,a)$ with endpoints q and a' for $a' \in \bar{U}_0(p) - U_0(p)$. If $a \in \alpha$ then $a = x$ or $a = x'$. In this case define $a(\Theta,q,\alpha) \in \bar{\alpha}(q,a)$ for $a(\Theta,q,\alpha)q = \Theta aq$. If we have $0 \le \Theta \le (aq)^{-1}(a'q)$, then $a(\Theta,q,\alpha)$ will exist. If $a \notin \alpha$ then let

$$a(\Theta,q,\alpha) \in \bar{Q}(x(\Theta,q,\alpha)\ ,\ x'(\Theta,q,\alpha)) \quad \text{and}$$

$$a(\Theta,q,\alpha) \in \bar{\alpha}(q,a) \quad \text{with} \quad a(\Theta,q,\alpha)q = \Theta aq.$$

From the definition, it is clear that $a(\Theta_1,q,\alpha)(\Theta_2,q,\alpha)$ equals $a(\Theta_1\Theta_2,q,\alpha)$ when these points exist.

Definition: The space R is called differentiable at p with respect to the segment α through p if the following condition holds

$$D_p: \lim_{\nu\to\infty} \frac{a_\nu(\Theta_\nu,p,\alpha)b_\nu(\Theta_\nu,p,\alpha)}{a_\nu b_\nu} = 1$$

where $a_\nu b_\nu > 0$, $0 < \Theta_\nu < M$ and $a_\nu, b_\nu \to p$.

(III.16) If D_p holds, there is a neighborhood $V(p)$ such that for $ab > 0$ in $V(p)$ and $0 < \Theta < 1$ then

$$\frac{1}{2} < \frac{a(\Theta,p,\alpha)\ b(\Theta,p,\alpha)}{\Theta ab} < \frac{3}{2}\ .$$

When p and α are understood, we set $a(\Theta) = a(\Theta,p,\alpha)$ and

$b(\Theta) = b(\Theta,p,\alpha)$.

(III.17) If D_p holds, then $m(a,b) = \lim_{\Theta\to 0+} \Theta^{-1} a(\Theta)b(\Theta)$ for $a,b \in V(p)$ exists, and

$$\lim_{\nu\to\infty} \frac{m(a_\nu,b_\nu)}{a_\nu b_\nu} = 1 \quad \text{for} \quad a_\nu,b_\nu \to p \quad \text{with} \quad a_\nu b_\nu > 0.$$

Proof: If $ab = 0$ then (III.16) gives that $a(\Theta)b(\Theta) = 0$ for all $0 < \Theta < 1$. Assume $ab > 0$ then (III.16) implies

$$0 < \frac{ab}{2} \le \lim\inf \Theta^{-1} a(\Theta)b(\Theta) \le \lim\sup \Theta^{-1} a(\Theta)b(\Theta) \le \frac{3ab}{2} .$$

Choose $\Theta_\nu \to 0^+$ such that $\lim \Theta_\nu^{-1} a(\Theta_\nu)b(\Theta_\nu)$ exists and let $\gamma_\nu \to 0^+$. Let $\Theta_{\nu'}$ be the smallest $\Theta_n > \gamma_\nu$ then $\Theta_{\nu'} \to 0$ and

$$\lim_{\nu\to\infty} \frac{a(\gamma_\nu)\ b(\gamma_\nu)}{(\gamma_\nu/\Theta_{\nu'})\ a(\Theta_{\nu'})\ b(\Theta_{\nu'})} = 1$$

Therefore, $m(a,b)$ exists and is different from 0 and ∞ if $ab > 0$.

If $a_\nu b_\nu > 0$ and $a_\nu,b_\nu \to p$, choose $\Theta_\nu < \nu^{-1}$ such that $|m(a_\nu,b_\nu) - \Theta_\nu^{-1} a_\nu(\Theta_\nu)\ b_\nu(\Theta_\nu)| < \nu^{-1} a_\nu b_\nu$ then

$$\frac{m(a_\nu,b_\nu)}{a_\nu b_\nu} \to 1.$$

(III.18) $m(a(\Theta),b(\Theta)) = \Theta m(a,b)$.

Proof: $m(a(\Theta),b(\Theta)) = \lim_{\gamma\to 0+} \gamma^{-1} a(\gamma\Theta)\ b(\gamma\Theta) = \Theta \lim_{\Theta\gamma\to 0+} (\Theta\gamma)^{-1} a(\Theta\gamma)\ b(\Theta\gamma)$.

The next few lemmas are concerned with the continuity of $m(a,b)$.

(III.19) Let $a_\nu \to a$, $b_\nu \to b$ with $a_\nu b_\nu > 0$. For fixed n there is a ν_n such that for all $\nu \ge \nu_n$,

$$|a_\nu(n^{-1})b_\nu(n^{-1}) - a(n^{-1})b(n^{-1})| < n^{-2}.$$

<u>Proof</u>: It is clear from the definition of $a(\Theta)$ that if $a_\nu \to a$, then $a_\nu(\Theta) \to a(\Theta)$.

(III.20) If $a_\nu \to a$, $b_\nu \to b$ then $m(a_\nu, b_\nu) \to m(a,b)$.

<u>Proof</u>: If $ab = 0$, then $a_\nu b_\nu \to 0$ and $m(a_\nu, b_\nu) \to 0 = m(a,b)$. Assume that $ab > 0$ and $m(a_\mu, b_\mu) \to \gamma \neq m(a,b)$ for some subsequence $\{\mu\} \subset \{\nu\}$. This cannot happen since (III.19) guarantees a subsequence $\{\mu_n\} \subset \{\mu\}$ exists such that

$$|n^{+1}a_{\mu_n}(n^{-1})b_{\mu_n}(n^{-1}) - n^{+1}a(n^{-1})b(n^{-1})| < n^{-1}$$

so that

$$n^{+1}a_{\mu_n}(n^{-1})b_{\mu_n}(n^{-1}) \to m(a,b)$$

yet

$$\frac{m(a_{\mu_n}, b_{\mu_n})}{n^{+1}a_{\mu_n}(n^{-1})b_{\mu_n}(n^{-1})} = \frac{m(a_{\mu_n}(n^{-1}), b_{\mu_n}(n^{-1}))}{a_{\mu_n}(n^{-1})b_{\mu_n}(n^{-1})} \to 1.$$

(III.21) Let $\Theta_\nu \to 0^+$, $a_\nu b_\nu > 0$, $a_\nu \to a$, $b_\nu \to b$, then

$$m(a,b) = \lim \Theta_\nu^{-1} a_\nu(\Theta_\nu) b_\nu(\Theta_\nu).$$

<u>Proof</u>: $m(a_\nu(\Theta_\nu), b_\nu(\Theta_\nu)) = \Theta_\nu m(a_\nu, b_\nu)$, yet

$$\frac{m(a_\nu, b_\nu)}{\Theta_\nu^{-1} a_\nu(\Theta_\nu) b_\nu(\Theta_\nu)} = \frac{\Theta_\nu m(a_\nu, b_\nu)}{a_\nu(\Theta_\nu) b_\nu(\Theta_\nu)} = \frac{m(a_\nu(\Theta_\nu), b_\nu(\Theta_\nu))}{a_\nu(\Theta_\nu) b_\nu(\Theta_\nu)} \to 1,$$

because of (III.20).

(III.22) Let $a_\nu, b_\nu \to p$ with $a_\nu b_\nu = 0$ and $\bar{\alpha}(p,a_\nu) \to \bar{\alpha}_0$, $\bar{\alpha}(p,b_\nu) \to \bar{\alpha}_0$ where $\bar{\alpha}_0$ is a segment. Then $(pa_\nu)(pb_\nu)^{-1} \to 1$.

Proof: Let $\bar{a}_\nu \in \bar{\alpha}(p,a_\nu)$ and $\bar{a}_\nu \to q \neq p$. Then $\bar{a}_\nu(\Theta_\nu^{-1}) = a$ for $\Theta_\nu \to 0^+$. Set $\bar{b} = b_\nu(\Theta_\nu^{-1})$ then $\bar{a}_\nu \bar{b}_\nu = 0$ and $\bar{b}_\nu \to q$.

This gives

$$\frac{\Theta_\nu m(p,\bar{a}_\nu)}{\Theta_\nu m(p,\bar{b}_\nu)} = \frac{m(p,a_\nu)}{m(p,b_\nu)} \to 1.$$

By (III.17) we have

$$\frac{m(p,a_\nu)}{pa_\nu} \to 1 \quad \text{and} \quad \frac{m(p,b_\nu)}{pb_\nu} \to 1.$$

Therefore,

$$(pa_\nu)(pb_\nu)^{-1} \to 1.$$

Since $ab = 0$ if, and only if, $m(a,b) = 0$ in $V(p)$ the original null Arcs remain null. We extend $V(p)$ to the tangent space T_p by the following method. If $a \in V(p)$ let $a(\sigma)$ be a point of T_p for $0 \leq \sigma < \infty$. If $a,b \in T_p$ define $m(a,b) = \gamma m(a(\gamma^{-1}),b(\gamma^{-1}))$ where $\gamma > 0$ and furthermore, $a(\gamma^{-1})b(\gamma^{-1}) \in V(p)$. The topology of T_p for $V(p)$ is the same as the original topology of $V(p)$. If $a \in T_p$ and $a \notin V(p)$ let a neighborhood $W(\gamma,U)$ of a for $\gamma > 0$, $a(\gamma^{-1}) \in V(p)$ and $U \subset V(p)$ with $a(\gamma^{-1}) \in U$ then consist of all b such that $b(\gamma^{-1}) \in U$.

The two partial orderings $<_1$ and $<_2$ on $V(p)$ can be extended to T_p by $a,b \in V(p)$ with $a <_i b$, then $a(\sigma) <_i b(\sigma)$ for $0 < \sigma < \infty$. If

$q,r,s \in T_p$ with $q <_i r <_i s$ then for sufficiently small positive Θ the points $q(\Theta), r(\Theta), s(\Theta)$ all lie in $V(p)$ and $q(\Theta)r(\Theta) + r(\Theta)s(\Theta) \le q(\Theta)s(\Theta)$. Multiplying through by Θ and taking the limit as $\Theta \to 0^+$, we find $m(q,r) + m(r,s) \le m(q,s)$.

If $a,b \in T_p$ with $m(a,b) > 0$, then a null box $Q(a,b)$ is determined in T_p. Clearly $\bar{Q}(a,b)$ is compact.

(III.23) Let $1 > \rho > 0$ and $m(a,b) > 0$. Then there exists a point c with $a <_i c <_i b$ and $m(a,c) = \rho m(a,b)$ as well as $m(a,c) + m(c,b) = m(a,b)$.

Proof: Let $0 < \Theta < 1$ and $d^\Theta \in \alpha(a(\Theta),b(\Theta)) \subset V(p)$ with $a(\Theta)d^\Theta$ equals $\rho a(\Theta)b(\Theta)$. Let $c^\Theta = d^\Theta(\Theta^{-1})$ then $a <_i c^\Theta <_i b$ and $c^\Theta \in Q(a,b)$. Since $\bar{Q}(a,b)$ is compact, the c^Θ have an accumulation point c. Let $c^{\Theta_\nu} \to c$ for $\Theta_\nu \to 0^+$. Then $m(a,c^\Theta) \to m(a,c)$. Yet,

$$\lim m(a,c^{\Theta_\nu}) = \lim \Theta_\nu^{-1} a(\Theta_\nu)d^{\Theta_\nu} = \rho \lim \Theta_\nu^{-1} a(\Theta_\nu)b(\Theta_\nu) = \rho\, m(a,b).$$

It follows by the same line of reasoning that

$m(b,c) = (1-\rho)m(a,b)$. Also, it is clear that $a <_i c <_i b$.

Regularity Condition:

R_p: Given $m(c,d) > 0$ in T_p with $c <_i d$ there is only one curve $x(t)$ for $0 \le t \le 1$ such that $x(t_1) <_i x(t_2)$ if $t_1 < t_2$ and such that $x(t)$ is a segment with respect to the metric $m(a,b)$ connecting c to d.

If $m(a,b) = 0$ then there is always a null curve in T_p from a to b obtained by letting $\alpha(a(\Theta),b(\Theta)) \subset V(p)$ and setting $\underline{\alpha}(a,b) = \{x \mid x(\Theta)$ belongs to $\alpha(a(\Theta),b(\Theta))\}$. For any $a,b \in T_p$ we let $\underline{\alpha}(a,b)$ be the (unique when R_p holds) segment from a to b in the sense of $m(a,b)$.

(III.24) Let R satisfy D_p and R_p. Let $d^\Theta \in \alpha(a(\Theta),b(\Theta))$ and $\lim_{\Theta\to 0^+} \frac{a(\Theta)d^\Theta}{a(\Theta)b(\Theta)} \to \rho$ for $m(a,b) > 0$ and $0 < \rho < 1$. Then $d^\Theta(\Theta^{-1}) \to c \in \underline{\alpha}(a,b)$ with $m(a,c) = \rho m(a,c)$.

Proof: Let a sequence $\Theta_\nu \to 0^+$ with $d^{\Theta_\nu}(\Theta_\nu^{-1})$ converging to some point $c \in \bar{Q}(a,b)$. Let $a(\Theta_\nu)d^{\Theta_\nu} = \rho_\nu a(\Theta_\nu)b(\Theta_\nu)$ then $m(a,d^{\Theta_\nu}(\Theta_\nu^{-1})) \to \rho m(a,b)$ and also $m(b,d^{\Theta_\nu}(\Theta_\nu^{-1})) \to (1-\rho)m(a,b)$ hence $c \in \underline{\alpha}(a,b)$ and furthermore $m(a,c) = \rho m(a,b)$.

It is now easy to show prolongation is possible in T_p.

(III.25) Given D_p and R_p any arc $\underline{\alpha}(a,b)$ may be prolonged in T_p.

Proof: If $m(a,b) = 0$ the result is clear. Assume $m(a,b) > 0$ and furthermore that $a,b \in V(p)$. Extend the original Arc $\alpha(a,b)$ to $\alpha(a,d)$ for $\bar{Q}(b,d) \subset V(p)$ and $aq > 0$ if $q \in \bar{Q}(b,d)$. If $\bar{Q}(b,d)$ has vertices b,e,d,f, then for $u \in \alpha(e,f)$ the new Arc $\underline{\alpha}(a,u)$ is unique and depends continuously on u. If $a <_i b$ then $a <_i u$. The point $v = \underline{\alpha}(a,u) \cap [\alpha(b,e) \cup \alpha(b,f)]$ moves continuously from e to f as u traverses $\alpha(e,f)$ and for some u_0 we have $b \in \underline{\alpha}(a,u_0)$. Thus, $\underline{\alpha}(a,u_0)$ extends $\underline{\alpha}(a,b)$.

Let now $p_\nu \to p$ and α_ν be a segment through p with endpoints on the boundary of $U_0(p)$ tending to a standard segment α through p. Using the same notation as before, we say the space is <u>continuously differentiable</u> at p if the following condition is satisfied

$$CD_p: \quad \lim \frac{a_\nu(\Theta_\nu,p,\alpha_\nu)\ b_\nu(\Theta_\nu,p,\alpha_\nu)}{\Theta_\nu a_\nu b_\nu} = 1$$

where $a_\nu, b_\nu \to p$, $0 < \Theta_\nu < M$, $a_\nu b_\nu > 0$.

(III.26) If a space R is continuously differentiable and regular at p then T_p is Minkowskian.

__Proof__: Let $m(a,b) > 0$, $m(a,c) > 0$ and $m(b,c) > 0$. Let $b' \in \underline{\alpha}(a,b)$ and $c' \in \underline{\alpha}(a,c)$ be the mid points of these Arcs with respect to the metric $m(x,y)$.

Let $d^\Theta \in \alpha(a(\Theta),b(\Theta))$ by $d^\Theta = b(\Theta)(\frac{1}{2},a(\Theta),\alpha_{a(\Theta)})$ where $\alpha_{a(\Theta)}$ is a segment through $a(\Theta)$ tending to α as $\Theta \to 0^+$. Let $e^\Theta = c(\Theta)(\frac{1}{2},a(\Theta),\alpha_{a(\Theta)})$. Then

$$\frac{a(\Theta)d^\Theta}{2^{-1}a(\Theta)b(\Theta)} \to 1 \quad \text{so that} \quad \Theta^{-1}a(\Theta)d^\Theta \to 2^{-1}m(a,b)$$

and

$$\frac{a(\Theta)e^\Theta}{2^{-1}a(\Theta)b(\Theta)} \to 1 \quad \text{so that} \quad \Theta^{-1}a(\Theta)e^\Theta \to 2^{-1}m(a,c)$$

Then $d^\Theta(\Theta^{-1}) \to b'$ and $e^\Theta(\Theta^{-1}) \to c'$. Also,

$$\frac{d^\Theta e^\Theta}{2^{-1}b(\Theta)e(\Theta)} \to 1 \quad \text{by } CD_p.$$

Therefore, $m(b',c') = \lim_{\Theta \to 0^+} \Theta^{-1}d^\Theta e^\Theta = 2^{-1}m(b,c)$.

This shows T_p has zero curvature. It follows from the definition of T_p and (III.10) that T_p is a doubly timelike Minkowskian plane.

CHAPTER IV

PLANES AND PATH-UNIQUE SPACES

1. Doubly Timelike Planes. A doubly timelike surface homeomorphic to A^2 is called a *doubly timelike plane*. It follows from Theorem (IV.1) that the universal covering space of a doubly timelike surface is a doubly timelike plane.

Let C be a curve in R given by $x(t)$ for $0 \leq t \leq 1$ with $x(0) = a$ and $x(1) = b$. Then we say C is an oriented curve from a to b. If ϕ is a homeomorphism of $[0,1]$ onto itself with $\phi(0) = 0$ and $\phi(1) = 1$ then $x(t), x(\phi(t))$ are called two representations of C. If C' is an oriented curve $x'(t)$ from b to c, then the product CC' of C and C' is defined as an oriented curve from a to c which may be represented by $f(t)$ such that for $0 \leq t \leq \frac{1}{2}$, $f(t) = x(2t)$ and for $\frac{1}{2} \leq t \leq 1$, $f(t) = x'(2t-1)$. The inverse C^{-1} to the oriented curve C is defined as the oriented curve from b to a given by $f(t) = x(1-t)$.

Let an oriented null element $n(a)$ be given at a point a on the curve C in the doubly timelike plane R. Then we can define uniquely an oriented null element $n(b)$ at each point b of C, called the translate of $n(a)$ along C from a to b and denoted $n(b) = n(a) \uparrow_c^b C$. If C' is a subcurve of C contained in a null box W, then for all x and y in C' the oriented null lines determined by $n(x)$ and $n(y)$ must belong to the same oriented family of null lines covering W simply.

It is obvious that $n(b) = n(a) \uparrow_a^b C$ implies $n(a) = n(b) \uparrow_b^a C^{-1}$. If $n'(a) \neq n(a)$ is a different oriented null element at a, then $n'(a) \uparrow_a^b C \neq n(a) \uparrow_a^b C$. If C is a curve joining a to b then we write $n(b) = n(a) \uparrow C$. If C is an oriented curve from a to b and C' an an oriented curve from b to c and C" another from c to d, then $n(a) \uparrow CC'C" = n(a) \uparrow C \uparrow C' \uparrow C"$.

(IV.1) <u>Theorem</u>. In a doubly timelike space R, let C_1 and C_2 be two homotopic oriented curves from a point p to a point q. Let n(p) be an oriented null element at p, then $n(p) \uparrow C_1 = n(p) \uparrow C_2$.

<u>Proof</u>: Let ϕ be a homotopy mapping from the square S: $0 \leq s \leq 1$, $0 \leq t \leq 1$ into R, such that for $0 \leq s \leq 1$, $\phi(s,0) = p$, $\phi(s,1) = q$, C_1 is represented by $\phi(0,t)$ and C_2 by $\phi(1,t)$ for $0 \leq t \leq 1$. Let n be such that each square $S_{ij}: \frac{i-1}{n} \leq s \leq \frac{i}{n}$, $\frac{j-1}{n} \leq t \leq \frac{j}{n}$ for each $i,j = 1,\cdots,n$ is mapped by ϕ into some null box in R. Let $k = (j-1)n + i$, then each integral value of k between 1 and n^2 corresponds uniquely to a pair of values for i,j. Thus S_{ij} can be denoted by S_k. Let P(k) be the perimeter of $\bigcup_{i=1}^{k} S_i$, regarded as an oriented Jordan curve from z to z, where z is the origin of the (s,t)-plane. Then $\phi P(k)$ is an oriented closed curve in R, beginning and ending at p. * If $k \geq n$, then $\phi P(k)$ passes through q, and $\phi P(n^2) = C_1 C_2^{-1}$.

Since S_1 is contained by some null box, it is obvious that $n(p) = n(p) \uparrow \phi P(1)$. We now assert that if $n(p) \uparrow \phi P(k-1) = n(p)$, then $n(p) \uparrow \phi P(k) = n(p)$. P(k-1) has a portion P_k in common with the perimeter of S_k, namely, either two adjacent sides of S_k or just one side of S_k. P_k is oriented by the orientation of P(k-1), so that it is an oriented curve from some point p_k to another p_k'. P(k-1) is then the product of three oriented curves, namely, P_{k-1} from z to p_k, P_k from p_k to p_k' and P_{k-1}' from p_k' to z. Thus $P(k-1) = P_{k-1}P_kP_{k-1}'$. Let P_k' be the

---

* Let ϕA denote $\phi(A)$.

other portion of the perimeter of S_k, regarded as an oriented curve from p_k to p_k'. Then $P(k) = P_{k-1}P_k' P_{k-1}'$. Thus $n(p) = n(p) \uparrow \phi P(k-1) = n(p) \uparrow \phi(P_{k-1}P_kP_{k-1}') = n(p) \uparrow \phi(P_{k-1}P_k) \uparrow \phi(P_{k-1}') = n(p) \uparrow \phi(P_{k-1}P_k) \uparrow \phi(P_k^{-1} P_k') \uparrow \phi P_{k-1}'$ because $\phi(P_k^{-1} P_k')$ is contained in some null box, so that for any oriented null element $n(\phi p_k')$ at $\phi p_k'$, $n(\phi p_k') \uparrow \phi(P_k^{-1} P_k') = n(p_k')$. Thus

$$n(p) = n(p) \uparrow \phi(P_{k-1}P_kP_k^{-1} P_k' P_{k-1}') = n(p) \uparrow \phi(P_{k-1}P_k' P_{k-1}') = n(p) \uparrow \phi P(k).$$

This proves the assertion.

Thus by mathematical induction, $n(p) = n(p) \uparrow \phi P(k)$ is true for all $k = 1,\cdots,n^2$, and $n(p) = n(p) \uparrow \phi P(n^2) = n(p) \uparrow (C_1C_2^{-1}) = n(p) \uparrow C_1 \uparrow C_2^{-1}$. Therefore $n(p) \uparrow C_2 = n(p) \uparrow C_1 \uparrow C_2^{-1} \uparrow C_2 = n(p) \uparrow C_1$, which proves the theorem.

(IV.2) In a doubly timelike plane, an oriented null element $n(p)$ at a point p cannot be translated along any closed curve beginning and ending at p into a different oriented element at p. If q is any point in R, then for all curves C passing through p and q, $n(p) \uparrow_p^q C$ coincide.

(IV.3) In a doubly timelike plane R, the null lines can be partitioned into two families each covering R simply.

<u>Proof</u>: Let $n(p)$ be an oriented null element at a point p. Let q be any point in R. Define $n(q) = n(p) \uparrow_p^q C$ for any curve passing through p and q. Define $N(q)$ as the null line containing $n(q)$. Let F_N be $\{N(q) | q \varepsilon R\}$. If q' is any point on $N(q)$, then $N(q') = N(q)$ because $n(q') = n(p) \uparrow_p^q C \uparrow_q^{q'} N(q) = n(q) \uparrow_q^{q'} N(q)$. Therefore F_N is a family of null lines

covering R simply.

If $m(p)$ is an oriented null element at p not contained in $N(p)$, then for all q in R, define $m(q) = m(p) \uparrow_p^q C$ and $M(q)$ the null line containing $m(q)$, and $F_M = \{M(q) \mid q \in R\}$. Then F_M is again a family of null lines covering R simply. It is obvious that F_M and F_N have empty intersections. Therefore at each point q, $M(q) \neq N(q)$. F_M and F_N thus exhaust all the null lines in R.

The next four lemmas follow from (IV.1) and (IV.3), see [22].

(IV.4) In a doubly timelike plane all null paths are simple.

(IV.5) Two different null lines in a doubly timelike plane intersect at most once.

(IV.6) A geodesic in a doubly timelike plane intersects a null line at most once.

(IV.7) A geodesic in a doubly timelike plane is simple and homeomorphic to an open interval of the real axis.

When there is exactly one path connecting two distinct points p,q in a doubly timelike plane we let $P(p,q)$ denote this path when it exists. Let $P(p,q)$ have a representation $x(t)$ for $t \in I$ with $x(\alpha) = p$, $x(\beta) = q$. Then $T(p,q)$ is the subcurve of $P(p,q)$ given by $x(t) \mid [\alpha,\beta]$.

Let E^2 denote the euclidean (x,y)-plane. We define two partial orderings $\leq_1$ and $\leq_2$ on E^2 as follows:

1.) $(x_1,y_1) \leq_1 (x_2,y_2)$ if $x_1 \leq x_2$ and $y_1 \leq y_2$

2.) $(x_1,y_1) \leq_2 (x_2,y_2)$ if $x_1 \leq x_2$ and $y_2 \leq y_1$.

Definition: A homeomorphism ϕ mapping a subset S of the doubly timelike space R into E^2 is called a grid-map if all null Arcs in S are mapped into segments parallel either to the x-axis or y-axis.

For two points $p,q \in S$ we say $p \leq_i q$ if $\phi(p) \leq_i \phi(q)$. A curve C: $x(t)$ for $a \leq t \leq b$ is ordered by $\leq_i$ if $t_1 < t_2$ implies $\phi(x(t_1)) \leq_i \phi(x(t_2))$.

Let ϕ: $S \to E^2$ be a grid-map and let $\bar{B} \subset S$.

(IV.8) If $\phi(B)$ is a region in E^2 with sides parallel to the axes, then $p,q \in B$ implies there is a path through p and q.

Proof: It is sufficient to prove: For $\bar{B}$ with vertices a,b,c,d named in cyclic order, the point a can be joined to any point x on the two opposite sides by a path T(a,x) in $\bar{B}$ from a to x.

Let ϕ be a grid-map of $\bar{B}$ onto E^2 such that $a \leq_1 b \leq_1 c$, $a \leq_1 d \leq_1 c$, $d \leq_2 a \leq_2 b$ and $d \leq_2 c \leq_2 b$. Let u,v be points on T(a,b) and T(a,d) respectively, different from a but so close to a that $\alpha(u,v)$ exists and lies in a convex neighborhood of a. Let x be a variable point on $\alpha(u,v)$. Then $\alpha(a,x)$ is unique. Let $P_B(a,x)$ denote the path in $\bar{B}$ determined by $\alpha(a,x)$. The curve $P_B(a,x)$ is ordered by $\leq_1$ because it cannot intersect any null path in B in more than one point. Cover B with a finite number of null boxes so that $P_B(a,x)$ intersects each in at most one Arc. Thus $P_B(a,x)$ must intersect $T(b,c) \cup T(c,d)$ at a point x'.

Using continuity of the Arcs and compactness of $\bar{B}$ it is not hard to show x' depends continuously on x. Now x' = b for x = u and x' = d if x = v, hence a can be joined by some $T(a,x') \subset \bar{B}$ to any point of $T(b,c) \cup T(c,d)$.

(IV.9) Theorem. (Time Inequality Theorem). Let S be an open, path convex subset of a doubly timelike surface R, such that there is a grid-map $\phi: S \to E^2$. Then an indefinite metric $x'y'_S$ may be defined on S such that $x'y' = x'y'_S$ if $x',y' \in S$ and $(x',y') \in \Pi_0$. Furthermore, for each triple of points x',y',z' in S with $x' \leq_i y' \leq_i z'$ we have $y'z'_S + x'y'_S \leq x'z'_S$, equality being true if, and only if, x',y' and z' lie on the same path in S.

Proof: Without loss of generality, assume $x' \leq_1 y' \leq_1 z'$ and that there is no path in S passing through all of them. We shall prove $x'y'_S + y'z'_S < x'z'_S$ where $x'y'_S$ denotes the length of the curve $T(x',y')$.

The curves $T(x',y'), T(y',z'), T(x',z')$ form a triangle bounding a closed region T. T can be contained in the interior I of some triangle bounding a closed region $T' \subset S$. T,T',I are all path-convex. We shall prove that if $p' <_1 q' <_1 r'$ are three points in I not lying on a path in S, then $p'q'_S + q'r'_S < p'r'_S$. Since T' is compact, there is a sufficiently small positive number k such that any euclidean square s with side-length $4k$ and sides parallel to the x- and y-axes, which intersects $\phi T'$, is contained in ϕS. Note that $\phi^{-1}(s)$ is a null box. Thus $\phi T'$ intersects and can be covered by a finite number of euclidean squares $S_{ij} = \text{def}\{(x,y) \mid (i-1)k \leq x \leq ik,\ (j-1)k \leq y \leq jk\}$ for integral values of i,j. For each of these S_{ij}, let $S'_{ij} = \text{def}\{(x,y) \mid (i - \frac{3}{2})k < x < (i + \frac{1}{2})k,\ (j - \frac{3}{2})k < y < (j + \frac{1}{2})k\}$. Then all these S_{ij} and S'_{ij} intersect $\phi T'$ and $\phi^{-1} S_{ij}$ $\phi^{-1}(S'_{ij})$ are null boxes in S. S_{ij} and the corresponding S'_{ij} are arranged in a finite number of rows and columns, where all S_{ij} belonging to the same row have the same j. Let the center of S_{ij} be (x_{ij}, y_{ij}).

Let $S_1, S_2, \cdots, S_n$ be a row of S_{ij} and $S_1', \cdots, S_n'$ the corresponding

S'_{ij}, such that if (x_i, y_i) is the center of S_i for each i, then $y_{i-1} = y_i$ and $x_{i-1} + k = x_i$ for $i = 2,3,\cdots,n$.

Consider $S'_1 \cup S'_2$. Let R' be the side $\{(x,y) \mid x = x_1 + k,\ y_1 - k < y < y_1 + k\}$ of S'_2. We shall for brevity denote each point in S and its image under ϕ by primed and unprimed letters. Thus $p' \varepsilon S$ if, and only if, $p \varepsilon \phi S$. If $p,q \varepsilon \phi S$, $T(p,q)$ and $P(p,q)$ shall denote $\phi T(p',q')$ and $\phi P(p',q')$, and pq shall mean $p'q'_S$.

Let p',q',r' be three points in $\phi^{-1}(S'_1 \cup S'_2)$ not lying on the same path in S. We shall prove that $pq + pr < pr$. This is trivial if p,q,r are all in S'_1 or all in S'_2. Thus it suffices to prove the inequality for $p \varepsilon S'_1 - S'_2$, $r \varepsilon S'_2 - S'_1$. If q does not lie on R', then either $T(p,q)$ or $T(q,r)$ intersects R' at some point w, thus either $qw + wr < qr$ or $pq + qw < pq$ respectively, so that it suffices to prove that $pw + wr < pr$, i.e., it suffices to prove that $pq + qr < pr$ for $q \varepsilon R'$.

Let $q_i = T(p,q) \cap L'$, $r_1 = T(q_1,r) \cap R'$,

$q_2 = T(p,r_1) \cap L'$, $r_2 = T(q_2,r) \cap R'$,

$q_3 = T(pr_2) \cap L'$, $r_3 = T(q_3,r) \cap R'$, etc., and

$q_0 = T(p,r) \cap L'$, $r_0 = T(p,r) \cap R'$.

Then all the q_i, r_i are distinct, and $q_0 <_1 q_i <_1 q_{i-1}$, $r_0 <_1 r_i <_1 r_{i-1}$, so that $\lim q_i = q_\infty$ and $\lim r_i = r_\infty$ exist. $T(q_\infty,r) = \lim T(q_i,r) \supset \lim T(r_i,r)$, so that $r_\infty \varepsilon T(q_\infty,r)$. Similarly $q_\infty \varepsilon T(p,r_\infty)$, so that p,q_∞,r_∞,r all lie on a path, namely $P(p,r)$, so that $q_\infty = q_0$, $r_\infty = r_0$.

If it were true that $pq + qr \geq pr$, then since $q_1q + qr < q_1r$, we have $pq_1 + q_1r < pr$; since $pq_1 + q_1r_1 < pr_1$, we have $pr_1 + r_1r < pr$ still more; $q_2r_1 + r_1r < q_2r$, gives $pq_2 + q_2r < pr$

$pq_2 + q_2r_2 < pr_2$, gives $pr_2 + r_2r < pr$

and so on. At the limit, we have $pq_0 + q_0r > pr$ and $pr_0 + r_0r > pr$ which are absurd.

We have proved $pq + qr < pr$ for $p <_1 q <_1 r$ in $S_1' \cup S_2'$, if p,q,r do not lie on a path. If we now let $S_1' \cup S_2'$ and S_3' take the roles of S_1' and S_2', we can prove that $pq + qr < pr$ for $p <_1 q <_1 r$ in $S_1' \cup S_2' \cup S_3'$ not lying on a path, and similarly in $S_1' \cup S_2' \cup S_3' \cup S_4'$ etc. until $S_1' \cup \cdots \cup S_n'$. In particular, the inequality $p'q' + q'r' < p'r'$ holds for $p' <_1 q' <_1 r'$ in $T' \cap \phi^{-1}(\bigcup_{i=1} S_1')$.

Let each row of S_{1j}' be denoted by R_m, $m = 1,\cdots,N$. $\phi^{-1}(R_m \cup R_{m+1})$ unfortunately may not be path-convex. However, $\phi^{-1}R_m \cap T'$, $\phi^{-1}R_{m+1} \cap T'$, and $\phi^{-1}(R_m \cup R_{m+1}) \cap T'$ are path-convex. Thus the above argument again may be applied with $R_m \cap T'$ and $R_{m+1} \cap T'$ taking the roles of S_1' and S_2', so that $p'q' + q'r' < p'r'$ for $p' <_1 q' <_1 r'$ not lying on a path in $T' \cap \phi^{-1}(R_1 \cup R_2)$, and similarly in $T' \cap \phi^{-1}(R_1 \cup R_2 \cup R_3)$, and so on until $T' \cap \phi^{-1}(\bigcup_{m=1}^{N} R_m) = T'$. In particular, we have $x'y' + y'z' < x'z'$.

2. <u>Path-Unique Doubly Timelike Surfaces</u>. A general G-space is called <u>path-unique</u> if for each pair x,y of distinct points there is a unique path $P(x,y)$ containing x and y. Here we are interested in doubly timelike surfaces. However, the first three lemmas of this section hold for any path-unique general G-space.

(IV.10) Every path in a path-unique space is simple.

<u>Proof</u>: Assume p is a multiple point of the path P. Then there are two distinct Arcs $\alpha(p,q)$ and $\alpha(p,r)$ that determine different line elements of P at p. If q and r are sufficiently close to p, the Arc $\alpha(q,r)$

exists with $(p,x) \varepsilon \Pi_0$ for all $x \varepsilon \alpha(q,r)$. If x_1 and x_2 are distinct points of $\alpha(q,r)$ the Arcs $\alpha(p,x_1)$ and $\alpha(p,x_2)$ determine different line elements of P at p. This implies P has uncountable multiplicity at p which is impossible, see [7, p. 229].

A simple path is either homeomorphic to the real axis or a circle. Consequently, $P(x,y)$ has at most two subarcs from x to y.

(IV.11) In a path-unique space, let $\{a_\nu\}$ be a sequence of points converging to a and $\{b_\nu\}$ a sequence converging to b. If $a_\nu \neq b_\nu$ and $a \neq b$, then $P(a_\nu,b_\nu) \to P(a,b)$.

<u>Proof</u>: To show $\lim\sup P(a_\nu,b_\nu) \subset P(a,b)$ let $p \varepsilon \lim\sup P(a_\nu,b_\nu)$. Then there is a subsequence $\{\mu\}$ of $\{\nu\}$ with $p_\mu \to p$. If $A(x,y)$ denotes some subarc of $P(x,y)$ from x to y, then (at least) one of the following must hold for ∞ many $\{\mu\}$:

$$1.)\quad p_\mu \varepsilon A(a_\mu,b_\mu)$$

$$2.)\quad a_\mu \varepsilon A(p_\mu,b_\mu)$$

$$3.)\quad b_\mu \varepsilon A(a_\mu,p_\mu).$$

Since the argument is the same in all cases, assume 1.) holds for ∞ many μ. Then there is a subsequence $\{i\}$ of $\{\mu\}$, a point c with $(a,c) \varepsilon \Pi$ and a sequence $c_i \varepsilon P(a_i,b_i)$ such that $c_i \to c$ and a_i,c_i,p_i,b_i lie in this order on $P(a_i,b_i)$. Let $A(a,b)$ be the subcurve of $P(a,b)$ from a to b containing c and let $A(a_i,b_i)$ be the subcurve of $P(a_i,b_i)$ from a_i to b_i containing c_i. Using A_3 and the compactness of $A(a,b)$ we find $\lim A(a_i,b_i) = A(a,b)$ and hence $p \varepsilon P(a,b)$.

The same type of arguments as in [7, p. 230] can be used to show

$P(a,b) \subset \liminf P(a_\nu,b_\nu)$ and hence $P(a,b) = \lim P(a_\nu,b_\nu)$.

(IV.12) In a path-unique space, if $P(a,b)$ is a circular path and $\{a_\nu\}$ is a sequence of points converging to a, and $\{b_\nu\}$ a sequence converging to b, then for all sufficiently large ν the path $P(a_\nu,b_\nu)$ are circular.

Proof: Cover $P(a,b)$ by a finite number of U_i with $U_i \times U_i \subset \Pi_0$. Suppose there is a subsequence $\{\mu\}$ of $\{\nu\}$ such that $P(a_\mu,b_\mu)$ are not circular. Eventually for some $k \in \{\mu\}$, $P(a_k,b_k)$ must intersect a U_i in (at least) two non-intersecting Arcs $\alpha(c_1,d_1)$ and $\alpha(c_2,d_2)$. Then for all $x \in \alpha(c_2,d_2)$ the Arcs $\alpha(c_1,x)$ exist, are unique, and are contained in $P(a_k,b_k)$. They intersect one another at c_1 only. Hence, $P(a_k,b_k)$ has uncountable multiplicity contradicting [7, p. 229].

(IV.13) In a path unique doubly timelike surface all paths passing through a point on a circular path that are sufficiently close to the circular path are circular with the same length. Furthermore, there are no circular null paths.

Proof: Let $P(a,b)$ be a closed path with $(a,b) \in \Pi$. There are points c and d with (cbd), $\alpha(c,d) \not\subset P(a,b)$ and such that for all $x \in \alpha(c,d)$ the path $P(a,x)$ is closed. Let S be the set $\bigcup_{x \in \alpha(c,d)} P(a,x)$.

Case 1. Suppose $P(a,x)$ is a geodesic for each $x \in \alpha(c,d)$. There is a homeomorphism of S onto a closed lune of a euclidean sphere, whose north and south poles are identified such that each $P(a,x)$ is mapped linearly onto a meridian on the lune. Thus the interior S^0 of S is homeomorphic to a plane. Let C be the open Arc $\alpha(c,d) - c - d$. Let x and x' be any two points on C. From the last chapter, any path in S^0 separates S^0 into two parts each having the path as boundary in S^0. Hence any path in S^0 that intersects $P(a,x)$ also intersects $P(a,x')$. Let u and u'

be two points in S^0 with $u \varepsilon P(a,x)$ and $u' \varepsilon P(a,x')$. The path $P(u,u')$ contains an Arc α through u that determines a path P' in S^0. It is not difficult to show that S^0 is path convex. Denote paths in S^0 by using primes.

There are two families F_N and F_M of null paths in S^0 each covering S^0 simply. We assert each family is one parametric. Consider F_N. Each of its members intersects some $P(a,x)$ for x in C, hence intersects $P(a,x)$ for all x on C. Let x,x' be two points on C, and y any point on $P(a,x)$. Then the null path in F_N through y intersects $P(a,x')$ at y'. $P'(y,y')$ can be covered by a finite number of null boxes $B_1, B_2, \cdots, B_k$ such that $y \varepsilon B_1$, $y' \varepsilon B_k$ and B_k is arbitrarily small. A point y^* on $P(a,x)$ sufficiently close to y implies $P'(y^*,y^{*\prime})$ intersects all $B_1, \cdots, B_k$ so that $y^{*\prime}$ is in B_k, i.e., $y^{*\prime}$ is arbitrarily close to y^*. Thus y' depends on y uniquely and continuously. Similarly, y depends on y' uniquely and continuously, showing that the mapping $y \to y'$ is a homeomorphism from $P(a,x)$ to $P(a,x')$. Consequently, F_N, and similarly F_M are one parametric families and there exists a grid-map mapping S^0 into a bounded region of the euclidean plane such that for each x in C, $P(a,x)$ is ordered by $<_1$. Any euclidean line parallel to a coordinate axis and intersecting one of $\phi(P(a,x))$ or $\phi(P(a,x'))$ also intersects the other. Hence all $P(a,x)$ have two common endpoints $a' <_1 a''$ on the boundary of $\phi(S^0)$.

We shall now show $P(a,x)$ and $P(a,x')$ have the same length. Without loss of generality assume $x' <_2 x$. Let $\{a_\nu\}$ be a sequence of points on $P(a,x)$ such that $\{\phi(a_\nu)\}$ converges to a', $\{b_n\}$ a sequence of points on $P(a,x)$ such that $\{\phi(b_n)\}$ converges to a''. The two null paths in S^0

through b_n must intersect $P(a,x')$ at c_n and d_n with $c_n <_1 d_n$. We then have $a_\nu <_1 c_n <_1 b_n <_1 d_n$. Hence $a_\nu c_n \le a_\nu b_n \le a_\nu d_n$. As $\nu \to \infty$, $a_\nu b_n \to ab_n$, while $a_\nu d_n \to ad_n$, $a_\nu c_n \to ac_n$. As $n \to \infty$, ac_n and ad_n converge to the length of $P(a,x')$ while ab_n converges to the length of $P(a,x)$. This proves the proposition for Case 1.

Case 2. Suppose $P(a,b)$ is a null path homeomorphic to a circle. Then the paths $P(a,x)$ close to $P(a,b)$ are also closed yet are not null. Hence the lengths of these geodesics $P(a,x)$ take at most two possible finite values, one for each pair of opposite quadrants at a. If $\{x_\nu\}$ is a sequence of points converging to b, then eventually $P(a,x_\nu)$ are closed and converge to $P(a,b)$ so that the lengths of $P(a,x_\nu)$ tend to zero, which is a contradiction.

A geodesic passing through a point p is said to pass through a quadrant Q at p if the geodesic contains an Arc $\alpha(p,q) \subset Q \cup p$. An immediate consequence of the last proposition is the next lemma.

(IV.14) In a geodesically complete path-unique doubly timelike surface R, the geodesics passing through a point p and a quadrant at p are either all closed with the same length or all open with infinite length.

(IV.15) In a geodesically complete path-unique doubly timelike surface R, there exists no point through which all geodesics are circular.

Proof: This rather lengthy proof is divided into three parts.

1.) Suppose there is a point a through which all geodesics are circular, then we assert all geodesics in R are circular.

First we show that if S is an open null box containing a, then any geodesic G intersecting S is circular. Let ϕ be a grid-map from S to E^2 such that $G \cap S$ is mapped on a curve ordered by $<_1$. Let $b \in G \cap S$. Then there exists a point $c \in S$ such that $\phi(b) <_1 \phi(c)$ and $\phi(a) <_1 \phi(c)$. Then $P(a,c)$ and $P(b,c)$ pass through the same quadrant at c. Since $P(a,c)$ is circular, lemma (IV.14) guarantees $P(b,c)$ is circular. Yet, G and $P(b,c)$ pass through the same quadrant at b so that G is also circular.

Let $T = \{x \in R|$ all geodesics through x are circular$\}$. The above shows T is open. A slight modification of the above argument shows the complement of T is also open. Hence, $T = R$.

2.) Let $\alpha(a,c)$ be a null Arc in S such that $\phi(a) \leq_1 \phi(c)$ and $\phi(\alpha(a,c))$ is parallel to the x-axis. Let $\alpha(d,b)$ be a segment in S containing c in its relative interior and suppose $d <_1 c <_1 b$ as well as $a <_2 d$. Let $P(a,b)$ have length γ and let $y(t)$ be its arc-length parametrization for $0 \leq t \leq \gamma$ with $y(0) = y(\gamma) = a$. Let $ab = t_1$ so that $y(t)$ restricted to $[0,t_1]$ represents $\alpha(a,b)$. The null paths in S through $y(t)$ for $0 \leq t \leq t_1$ whose images under ϕ are parallel to the y-axis intersect $P_S(a,d)$ uniquely at points $y'(t)$ which depend continuously and uniquely on t. Let $K(t_1)$ be the set of null Arcs $\alpha(y(t),y'(t))$ for $0 \leq t \leq t_1$.

There exists a maximal $t_0 > t_1$ such that the following properties are true: For each positive $t < t_0$ there exists a unique $y'(t)$ on $P(a,d)$ depending continuously on t such that $P(y(t),y'(t))$ is a null path and

(A) $M(t)$ is a simple Jordan arc contained in $P(y(t),y'(t))$ with endpoints $y(t)$ and $y'(t)$

(B) If $t_2 < t_3 < t_0$, then $M(t_2) \cap M(t_3)$ is empty

(C) If $t < t_1$ then $M(t) \subset K(t_1)$

(D) The set $K(t_0)$ of points on $M(t)$ for $0 < t < t_0$ is homeomorphic to $\{(x,y) \mid x,y \geq 0, \ x + y < 1 \text{ and } x^2 + y^2 > 0\}$ in E^2

(E) If $t_2 < t_3 < t_0$ there is a null path N through $y(t_2)$ such that N has as subset a simple Jordan arc in $K(t_0)$ intersecting $M(t_3)$.

We assert that $t_0 = \gamma$ and $y'(t) \to a$ as $t \to \gamma$. Then the paths $M(t)$ for $0 \leq t \leq \gamma$ would bridge up $P(a,b)$ and $P(a,d)$ forming a set R' with interior homeomorphic to an open disk. The null path $P(a,c)$ in R' has endpoints on $P(a,b) \cup P(a,d)$. But, then a is the only possible endpoint and $P(a,c)$ is circular contradicting (IV.13).

3.) To show $t_0 = \gamma$, assume $t_0 < \gamma$. Let $y(t_0)$ be contained in some null box S^*. There exists $t_2 < t_0$ such that $M(t) \cap S^*$ is non-empty for $t_2 \leq t < t_0$ and $M(t) \cap S^*$ always belongs to one, say F_M, of the two families F_M and F_N of null lines in S^*. The geodesic $P(a,b)$ separates S^* into two parts S_1^* and S_2^*. For $t_2 \leq t < t_0$, $M(t) \cap S^*$ all lie in one of the two parts, say S_2^*. The null path $P(t_0)$ passing through $y(t_0)$ with $P(t_0) \cap S^* \in F_M$ and the null line N in F_N through $y(t_2)$ intersect at a point y_1 of S_2^* because of (E). Let y_2 be a point of $S_2^* \cap P(t_0)$ with $(y(t_0) y_2 y_1)$. Let $y_3 \in S_2^* \cap M(t_2)$ be such that $y(t_2), y_1, y_2, y_3$ form the vertices of a null box. Let $z \in \alpha(y_2, y_3)$. If z is close to y_3 the geodesic $P(y(t_2),t)$ intersects $P(a,d)$ because of (D). For all $z \in \alpha(y_2,y_3)$ the geodesics $P(y(t_2),z)$ are circular, pass through the same quadrant at $y(t_2)$ and do not pass through a. Therefore, they all intersect $P(a,d)$. Let $P(y(t_2),z) \cap P(a,d) = p(z)$.

Let $T(y(t_2), p(z))$ represent the simple Jordan arc on $P(y(t_2), p(z))$

containing z and with endpoints $y(t_2)$ and $p(z)$. These $T(y(t_2), p(z))$ form a set T homeomorphic to a closed triangle, bounded by the sides $T_1 = M(t_2)$, $T_2 = T(y(t_2), p(y_2))$ and T_3 a subcurve of $P(a,d)$. Now $P(t_0)$ intersects T_2 at y_2 so it must intersect $T_1 \cup T_3$ at another point. It cannot intersect T_1 because inside T the null paths are partitionable into two families each covering T simply and $M(t_2)$ and $P(t_0)$ lie on null paths in the same families. Consequently, $P(t_0)$ intersects T_3 at some point $y'(t_0)$. This means $M(t_0) \subset P(t_0)$ connecting $y(t_0)$ and $y'(t_0)$ exists. Covering $M(t_0)$ with a finite number of null boxes shows $M(t_0) = \lim_{t \to t_0^-} M(t)$ and for some $\varepsilon > 0$, $y'(t)$ exist for $t < t_0 + \varepsilon$, such that (A) through (E) are true. Thus the maximality of t_0 is contradicted.

If $y'(t) \in P(a,d)$ tends to some point $a' \neq a$ as $t \to \gamma^-$, then $\lim_{t \to \gamma} P(y(t), y'(t)) = P(a,d)$ is a null line which contradicts $a <_1 d$. Therefore, $y'(t) \to a$ as $t \to \gamma^-$. We have now proved the following theorem.

(IV.16) Theorem. Any geodesically complete path-unique doubly timelike space is one of the following two types:

1.) There is a number L and a pair of opposed quadrants at each point through which all geodesics are circular with length L. All other geodesics in the space are isometric to the whole real line and all null lines are open. This space is homeomorphic to the Möbius strip.

2.) Every path is open and every geodesic isometric to the whole real line. This type of space will be called a straight plane.

It follows, (compare [3, p. 81]) that:

(IV.17) In a doubly timelike surface of type (1) above, any geodesic homeomorphic to a circle intersects every other path.

CHAPTER V

STRAIGHT PLANES

1. The Grid-Representation of a Straight Plane. Let R be a straight plane. Because of theorem (IV.16), we may assume without loss of generality that $\Pi_0 = R \times R$.

Let E^2 denote the euclidean (x,y)-plane. A grid-line L(a,b) [or grid-segment S(a,b)] in E^2 is a euclidean line passing through the points a,b [or a euclidean segment with a,b as endpoints] that is parallel to either the x-axis or the y-axis. A grid-rectangle abcd is a euclidean rectangle in E^2 with vertices a,b,c,d whose sides S(a,b), S(b,c), S(c,d) and S(d,a) are grid-segments. A grid-path is a polygonal path in E^2 made up of a finite number of grid-segments. The coordinates of a point p are referred to as (x(p), y(p)).

Let R be a straight plane. A grid S in R is a connected open subset of R for which there exists a grid-map $\phi: S \to E^2$, called a grid-representation of S, such that:

1.) For some positive $\xi < 1$ the image $\phi(S)$ is contained in the euclidean square with vertices $(\pm\xi, \pm\xi)$.

2.) The intersection of a grid-line with $\phi(S)$ is either empty or homeomorphic to an open interval.

We call the next result the Grid-Representation Theorem. A proof can be found in Woo, [22].

(V.1) Theorem. Any straight plane is a grid.

This theorem is used throughout the rest of this chapter. Much of the development of this chapter is analogous to that in G-spaces. We show in the last section that a Desarguesian straight plane in which the parallel axiom holds and all circles are concave is Minkowskian.

2. Perpendiculars and Asymptotes in a Straight Plane. Let ϕ be a grid-representation of a straight plane R, and $\leq_1$ and $\leq_2$ the associated partial orderings in R as well as in E^2. If p is a point in R, then the two null lines through p separate R into 4 quadrants at p, which we shall denote by $S_i(p)$, $i = 1,2,3,4$, such that $S_3(p) <_1 p <_1 S_1(p)$ and $S_2(p) <_2 p <_2 S_4(p)$.

(V.2) In a straight plane R, if $\underline{g}$ is a geodesic ordered by $<_1$ not passing through a point p, then exactly one of $S_2(p)$ and $S_4(p)$ intersects $\underline{g}$.

Proof: If $\underline{g}$ intersects both $S_2(p)$ and $S_4(p)$ and q_2, q_4 are any two points on $\underline{g} \cap S_2(p)$, $\underline{g} \cap S_4(p)$, then $q_2 <_2 q_4$, contradicting the assumption that $\underline{g}$ is ordered by $<_1$. On the other hand if neither $S_2(p)$ nor $S_4(p)$ intersects $\underline{g}$, then $\underline{g}$ is contained in either $S_1(p)$ or $S_3(p)$. Assume without loss of generality that $\underline{g} \subset S_1(p)$. Let $\{q_i\}$ be a sequence of points on $\underline{g}$ with $q_{i+1} <_1 q_i$ and $q_1 q_i \to \infty$ as $i \to \infty$. Then $pq_1 > pq_i + q_i q_1$ for all i, so that pq_1 is infinite, which is absurd.

Let us assume without loss of generality that $\underline{g}$ intersects $S_2(p)$, then a point f on $\underline{g} \cap S_2(p)$ is called the foot of p on $\underline{g}$ if $pf \geq px$ for all x on $\underline{g} \cap S_2(p)$. It may not exist, unless $\underline{g} \cap S_2(p)$ has compact closure in R, i.e., $\underline{g}$ intersects both null lines through p.

Let F_M, F_N denote the two families of null lines mapped by ϕ into euclidean lines parallel to the x- and y-axis respectively. A member of F_M, F_N passing through a point p will be denoted by $M(p)$ or $N(p)$ respectively.

(V.3) Let f be the foot of a point p on a geodesic $\underline{g}$ not passing through p, and q be any point on $\alpha(p,f) - p$. Then f is the unique foot of q on $\underline{g}$.

Proof: Assume ϕ is so chosen that $\underline{g}$ is ordered by $<_1$, and $\underline{g} \cap S_2(p)$ is empty. Let x be any point on $\underline{g} \cap S_4(q)$, then $p <_2 q <_2 x$, so that $px > pq + qx$. Hence $qx < px - pq \leq pf - pq = fq$.

Definition: If $\underline{g}$ is any geodesic, then we shall define $R(\underline{g})$ to equal $\{x \in R \mid x \leq_1 p$ and $q \leq_1 x$ for some p,q on $\underline{g}\}$ or $\{x \in R \mid p \leq_2 x \leq_2 q$ for some p,q on $\underline{g}\}$ depending upon whether $\underline{g}$ is ordered by $\leq_1$ or $\leq_2$. $R(\underline{g})$ is an open set, and if $p \in R(\underline{g})$, then p always has a foot on $\underline{g}$ because the two null lines through p both intersect $\underline{g}$.

(V.4) Theorem. Let $\underline{g}$ be a geodesic not passing through a point p. If f is a foot of p on $\underline{g}$ and each point in $R(\underline{g})$ has a unique foot on $\underline{g}$, then all points on $P(p,f) \cap R(\underline{g})$ have f as the only foot on $\underline{g}$.

Proof: Without loss of generality assume $\underline{g}$ is ordered by $\leq_1$ and $\underline{g} \cap S_2(p)$ is empty. Let q be any point on $P(p,f) \cap R(\underline{g})$ on the side of $\underline{g}$ not containing p. We shall prove that f is the foot of q on $\underline{g}$. The two null lines through q intersect $\underline{g}$ at p_1, p_2 where $p_1 <_1 p_2$. Any point z on $\alpha(p_1,q) \cup \alpha(p_2,q)$ has a unique foot $f(z)$ on $\underline{g}$, which depends on z continuously due to uniqueness. When $z = p_1$ or p_2, $f(z)$ coincides with z, hence there is a point q' on $\alpha(p_1 q) \cup \alpha(p_2,q)$ having f as foot on $\underline{g}$. We assert that $q' = q$. For if $q' \neq q$, then $f' = \alpha(p,q') \cap \underline{g} \neq f$. Now $fq' > f'q'$, $fp \geq f'p$, hence $pf + fq' > pq'$, contradicting the time-inequality. Thus $q = q'$ and f is the foot of all points on $P(p,f) \cap R(\underline{g})$ on the side of $\underline{g}$ not containing p. Let q now take the role of p, then the same argument yields f as the unique foot of all points on $P(p,f) \cap R(\underline{g})$ on the same side of $\underline{g}$ as p. The theorem is now proved.

Definition: A geodesic $\underline{h}$ is perpendicular to another geodesic $\underline{g}$, and $\underline{g}$ is transversal to $\underline{h}$ if a point $\underline{h} \cap \underline{g}$ exists and is the foot of all points of $\underline{h} \cap R(\underline{g})$ on $\underline{g}$. [Compare with definition p. 23.]

(V.5) If each point of $R(\underline{g})$ has a unique foot on $\underline{g}$, then through each point in $R(\underline{g}) \cap \underline{g}$ there is a unique perpendicular to $\underline{g}$.

Definition: $K'(p,q)$ will denote a metric circle passing through a point q and having p as center. It always has 4 components $K_i(p,q)$, $i = 1,2,3,4$, such that for each i, $K_i(p,q) \subset S_i(p)$. The particular component passing through q will be denoted by $K(p,q)$. $U(p,q)$ will denote the interior $\{p' \mid pp' < pq\}$ of $K'(p,q)$. $R - U(p,q) - K'(p,q)$ has four components, namely $S_i(p) - \bar{U}(p,q)$ for $i = 1,2,3,4$ where $\bar{U}(p,q) = U(p,q) \cup K'(p,q)$ is the closure of U. A component $K_i(p,q)$ of $K'(p,q)$ is concave if $S_i(p) - \bar{U}(p,q)$ is path-convex. It is strictly concave if any segment $\alpha(r,s)$ with r,s in the closure of $S_i(p) - \bar{U}(p,q)$ has its interior points in $S_i(p) - \bar{U}(p,q)$. If $K_i(p,q)$ is (strictly) concave for $i = 1,2,3,4$, then we say $K'(p,q)$ and $U(p,q)$ are (strictly) concave.

(V.6) Theorem. The following statements are equivalent in a straight plane R:

A.) All circles are concave.

B.) All circles are strictly concave.

C.) For any point p and a segment α contained in a quadrant of p, there is a unique point f on α such that $pf \geq px$ for all x on α, which we shall call the foot of p on α.

D.) If a segment α lies inside a quadrant of a point p, α being parametrized by its arc-length t as $x(t)$, then $px(t)$ regarded as a function of t is strictly troughless in the following sense: a function $f(t)$ is <u>strictly troughless</u> if $-f(t)$ is strictly peakless as defined in Busemann [3] 18.11.

<u>Proof</u>: (A) $\Rightarrow$ (B): Let $K(p,q)$ be a component of the circle $K'(p,q)$. Let ϕ be so chosen that $K(p,q) \subset S_3(p)$. Let r,s be two points in $S_3(p) - U(p,q)$. Let t be an interior point of $\alpha(r,s)$. We shall prove $pt > pq$. Let p' be any point on $P(t,p)$ such that $p <_1 p'$. Then $p't \geq$ either $p's$ or $p'r$ because $K(p',t)$ is concave. Without loss of generality, assume $p't \geq p'r$. Then $pr + pp' < p'r \leq p't$, so that $pt = p't - pp'$ which implies $pt > pr \geq pq$.

(B) $\Rightarrow$ (C) is obvious.

(C) $\Rightarrow$ (D): Let ϕ be so chosen that $\alpha \subset S_4(p)$. (D) is trivial if α is ordered by $<_2$. Suppose α is ordered by $<_1$ and (D) is false. Then there are three points $a <_1 b <_2 c$ on α such that $pa > pb < pc$. Let m be any number such that $pb < m < \min\{pa,pc\}$. Let a' be the closest point to b among all points on $\alpha(a,b)$ with $pa' = m$, and c' the closest point to b among points of $\alpha(b,c)$ with $pc' = m$. Then a' and c' are feet of p on $\alpha(a',c')$, contradicting (C).

(D) $\Rightarrow$ (A) is obvious.

<u>Definitions</u>: Let $\underset{\rightarrow}{r}$ be a geodesic ray lying on a geodesic $\underline{r}^+$ with the same orientation. $\underset{\rightarrow}{r}$ and $\underline{r}^+$ are said to be <u>ordered by</u> $<_1$ if for any two points p and q on $\underline{r}$ with p preceding q, we have $p <_1 q$. $\underset{\rightarrow}{r}$ and $\underline{r}^+$ are <u>ordered inversely by</u> $<_1$ if p preceding q on $\underline{r}$ implies $q <_1 p$. Similarly we can define the meaning of $\underset{\rightarrow}{r}$ and $\underline{r}^+$ being ordered or ordered inversely by $<_2$. A sequence of points $\{x_i\}$ on $\underline{r}^+$ is said to $\to \infty$ on $\underset{\rightarrow}{r}$ or along $\underline{r}^+$ if each point on $\underset{\rightarrow}{r}$ eventually precedes x_i. The points $\{x_i\} \to \infty$ monotonically on $\underline{r}$ or along $\underline{r}^+$ if x_i precedes x_{i+1} for all i. Let $\{p_i\}$ be a sequence of points converging to a point p in R. Then the following properties have been proved in Busemann [3] 11.2: The oriented segments $\alpha^+(p_i,x_i)$ converge in Hausdorff's sense to a ray $\underset{\rightarrow}{A}(p,\underset{\rightarrow}{r})$ with p as initial point, which we shall call the <u>asymptotic ray</u> through p to $\underset{\rightarrow}{r}$. It lies in $\underline{r}^+$ or does not intersect $\underline{r}^+$ according to whether p lies on $\underline{r}^+$ or not. The path $\underline{A}^+(p,\underset{\rightarrow}{r})$ containing $\underset{\rightarrow}{A}(p,\underset{\rightarrow}{r})$ with the same orientation is called an <u>asymptote</u> or <u>asymptotic line</u> through p to $\underset{\rightarrow}{r}$. If p' is a point on $\underline{A}^+(p,\underset{\rightarrow}{r})$, then $\underline{A}^+(p',\underset{\rightarrow}{r}) = \underline{A}^+(p,\underset{\rightarrow}{r})$ and $\underset{\rightarrow}{A}(p,\underset{\rightarrow}{r})$ contains or is contained by $\underset{\rightarrow}{A}(p',\underset{\rightarrow}{r})$ according to whether p precedes p' or p' precedes p on $\underline{A}^+(p,\underset{\rightarrow}{r})$. Thus two rays with the same initial point cannot be asymptotic to each other or to a third ray. Moreover, the set of all asymptotic lines to a ray covers R simply. $\underset{\rightarrow}{A}(p,\underset{\rightarrow}{r})$ and $\underline{A}^+(p,\underset{\rightarrow}{r})$ may also be denoted by $\underset{\rightarrow}{A}(p,\underline{r}^+)$ and $\underline{A}^+(p,\underline{r}^+)$ respectively, and $R(\underset{\rightarrow}{r})$ will mean $R(\underline{r}^+)$.

(V.7) If $p \in R(\underset{\rightarrow}{r})$, then $\underset{\rightarrow}{A}(p,\underset{\rightarrow}{r})$ is not a null line.

<u>Proof</u>: Suppose $\underset{\rightarrow}{A}(p,\underset{\rightarrow}{r})$ is a null line, then it intersects $\underline{r}$, which is absurd.

<u>Definition</u>: $R'(\underset{\rightarrow}{r}) = R'(\underline{r}^+)$ is defined as $\{x \mid x <_1 p$ for some p on $\underline{r}^+\}$

where $<_1$ is associated with some grid-representation of R such that $\underline{r}^+$ is ordered by $<_1$.

By (V.7) we know that a ray asymptotic to $\underline{r}^+$ is not null if it intersects $R(\underrightarrow{r})$. However, (V.7) does not imply $\underline{r}^+$ has no null asymptotes.

(V.8) If $\underline{r}^+$ is ordered by $<_1$, then $R'(\underrightarrow{r}) \supset R(\underrightarrow{r})$ and the boundary of $R'(\underrightarrow{r})$ in each side of $\underline{r}^+$ is either empty or a null line asymptotic to $\underline{r}^+$, whose image under ϕ has an endpoint same as that of $\phi\underline{r}$ on the boundary of ϕR. If a point $p \in R'(\underline{r})$, then $\underline{A}^+(p,\underrightarrow{r}) \subset R'(\underline{r})$, and among the two null rays initiating from p and ordered by $<_1$, one intersects $\underline{r}^+$ and cannot be $\underrightarrow{A}(p,\underrightarrow{r})$, and if the other is not completely inside $R'(\underrightarrow{r})$, then $\underrightarrow{A}(p,\underrightarrow{r})$ is a geodesic ray.

Proof: These are easily verified by the grid-representation theorem.

3. Limit Circles in a Straight Plane. Although a sequence of circles will in general not converge, we do have the following result.

(V.9) Let $\underrightarrow{r}$ be a geodesic ray passing through a point y in a straight plane R. If $\{x_i\}$ is a sequence of points $\to \infty$ on $\underrightarrow{r}$, then $\lim_{i\to\infty} K(x_i,y)$ exists, and is a curve which will be called the limit circle $K_\infty(\underrightarrow{r},y)$ or $K_\infty(\underline{r}^+,y)$ through y to $\underrightarrow{r}$ and $\underline{r}^+$.

Proof: Let the grid-representation ϕ be so chosen that $\underrightarrow{r}$ is ordered by $<_1$. Let $\underrightarrow{M}$ and $\underrightarrow{N}$ be the two null rays initiating from y and ordered by $\leq_1$. Let x,x' be two points on $\underrightarrow{r}$ with $y <_1 x <_1 x'$. Let $K = K(x,y)$, $K' = K(x',y)$. We assert that $K \cap K' = y$ and $K' \cap S_3(x) \subset y \cup U(x,y)$. For if z is a point on K different from y, then $z <_1 x <_1 x'$, so that $x'z > xx' + xz = xx' + xy = x'y$. Thus z lies outside $U(x',y)$ and K' intersects $\alpha(x',z)$ at some point $z' \neq z$ in $S_2(y)$ or $S_4(y)$. Thus $K \cap K'$ contains y only, and $K' \cap (S_3(x) - y) \subset U(x,y)$.

Let $K_i = K(x_i,y)$ and assume without loss of generality that $x_i <_1 x_{i+1}$ for all i. Then ϕK_i are curves in E^2 ordered by $<_2$ passing through ϕy. Let F be the family of euclidean lines ordered by $<_1$ intersecting the coordinate axes at an angle $\pi/4$, and intersecting $\phi \underset{\rightarrow}{M}$ and $\phi \underset{\rightarrow}{N}$; then ϕK_i intersect each member of F. Let g be a member of F, then $y_i = g \cap \phi K_i$ satisfies $y_i <_1 y_{i+1}$ for all i. Let g intersect $\phi(\underset{\rightarrow}{M} \cup \underset{\rightarrow}{N})$ at y_0, then $y_i <_1 y_0$ for all i. Hence $y_\infty = \lim_{i\to\infty} y_i$ exists on $g \cap \phi R$, and obviously y_∞ depends on g continuously, as $y_\infty(g)$. The set of all $y_\infty(g)$ for g in F constitutes a curve ordered by $<_2$, whose inverse image under ϕ is $K_\infty(\underset{\rightarrow}{r},y)$.

(V.10) $K_\infty(\underset{\rightarrow}{r},y) \subset R'(\underset{\rightarrow}{r})$, and if $\underline{k}^+$ is a geodesic asymptotic line to $\underset{\rightarrow}{r}$ in $R'(\underset{\rightarrow}{r})$, then $K_\infty(\underset{\rightarrow}{r},y) \cap \underline{k}^+$ contains at most one point.

Definition: Let $\underset{\rightarrow}{r}$ be a geodesic ray through a point y, and x_i be a sequence of points on $\underset{\rightarrow}{r} \to \infty$ monotonically. Assume ϕ is so chosen that $\underset{\rightarrow}{r}$ is ordered by $<_1$. Then $K(x_i,y)$ separates R into two parts, namely, $S_3(x_i) - K(x_i,y) - U(x_i,y)$ and $(R - S_3(x_i)) \cup U(x_i,y)$, which will be called the exterior $E(x_i,y)$ and interior $I(x_i,y)$ of $K(x_i,y)$. The above theorem shows that $K_\infty(\underset{\rightarrow}{r},y)$ separates R into two parts $\bigcup_i E(x_i,y)$ and the interior of $(\bigcap_i I(x_i,y))$, which will be called the exterior $E_\infty(\underset{\rightarrow}{r},y)$ and interior $I_\infty(\underset{\rightarrow}{r},y)$ of $K_\infty(\underset{\rightarrow}{r},y)$ respectively.

(V.11) Let $\underset{\rightarrow}{r}$ be a geodesic ray passing through a point y. Let $\underline{g}^+$ be any asymptotic line to $\underset{\to}{r}$ in $R'(\underset{\to}{r})$ such that $y \in R'(\underline{g}^+)$. Then $K_\infty(\underset{\to}{r},y)$ intersects $\underline{g}^+$ at a single point.

Proof: Let ϕ be so chosen that $\underset{\to}{r}$ is ordered by $<_1$. Then so is $\underline{g}^+$. Let x be a point on $\underset{\to}{r}$ such that $y <_1 x$ and $z_1 <_1 x$ for some z_1 on $\underline{g}^+$. Let z_i be a sequence of points $\to \infty$ monotonically along $\underline{g}^-$. Then $xz_i > z_1z_i + xz_1$ for all $i > 1$. As $i \to \infty$, $z_1z_i \to \infty$, so that $xz_i \to \infty$. By (V.2) $\underline{g}^+$ intersects both $S_1(x)$ and $S_3(x)$, so that for a variable point $z <_1 x$ on $\underline{g}^+$, xz is a continuous function of z taking all non-negative values exactly once. Thus there is a unique point $z(x)$ depending on x, $z(x) <_1 x$, which lies on $\underline{g}^+$ such that $xz(x) = xy$. Then $K(x,y)$ passes through $z(x)$.

Now $y \in R'(\underline{g}^+)$ implies that $\underline{g}^+$ intersects $R(\underset{\to}{r})$, so that $\underline{g}^+$ is a geodesic. By (V.2) $\underline{g}^+$ must intersect both $S_1(y)$ and $S_3(y)$, so that $\underline{g}^+$ intersects the boundary of $S_1(y)$ at some point z_0. If $y <_1 x_i$ is a sequence of points $\to \infty$ monotonically along $\underset{\to}{r}$, then $z(x_i) <_1 z(x_{i+1}) <_1 z_0$ for all i. Hence $z_\infty = \lim_{i\to\infty} z(x_i)$ exists, and lies on $K_\infty(\underset{\to}{r},y)$.

(V.12) Let $K_1 = K_\infty(\underset{\to}{r},y_1)$, $K_2 = K_\infty(\underset{\to}{r},y_2)$ be two limit circles with respect to a geodesic ray $\underset{\to}{r}$ through two points y_1,y_2 on $\underset{\to}{r}$. An asymptotic line $\underline{g}^+$ to $\underset{\to}{r}$ in $R'(\underline{r})$ does not intersect any limit circle to $\underset{\to}{r}$ if $\underline{g}^+$ is a null line. If $\underline{g}^+$ is a geodesic, and intersects some limit circle to $\underset{\to}{r}$, then it intersects all limit circles to $\underset{\to}{r}$, and if $z_i \in \underline{g}^+ \cap K_i$ for $i = 1$, 2, then $z_1z_2 = y_1y_2$.

Proof: Let $\underline{r}^+$ be the oriented geodesic containing $\underset{\to}{r}$ with the same orientation. Let ϕ be so chosen that $\underset{\to}{r}$ is ordered by $<_1$. Suppose $\underline{g}^+$ is a geodesic asymptote to $\underset{\to}{r}$ in $R'(\underline{r}^+)$ for which there is some y_0 on $\underline{r}^+$

such that $\underline{g}^+ \cap K_\infty(\underset{\to}{r}, y_0) = z_0$ exists. Let y_0' be any other point on $\underline{r}^+$. We shall prove $\underline{g}^+ \cap K_\infty(r, y_0')$ exists.

Let x_i be a sequence of points $\to \infty$ monotonically along $\underline{r}^+$. Eventually z_0 and $y_0' <_1 x_i$. Let S be a small neighborhood of z_0. Eventually $K(x_i, y_0)$ intersects S and so intersects $\underline{g}^+$ at a single point z_i which $\to z_0$ as $i \to \infty$. Let i_0 be some number such that for all $i > i_0$, z_i exist, $z_i <_1 y_0' <_1 x_i$. Since $P(x_i, z_i)$ has infinite length, for each $i > i_0$, there is a point $z_i' <_1 x_i$ on $P(x_i, z_i)$ such that $x_i z_i' = x_i y_0'$. Thus $K(x_i, y_0')$ intersect $P(x_i, z_i)$ at z_i'. Let z_0' be a point on $\underline{g}^+$ such that $z_0 z_0' = y_0 y_0'$ and that $z_0 <_1 z_0'$ or $z_0' <_1 z_0$ according as to whether $y_0 <_1 y_0'$ or $y_0' <_1 y_0$. Then $z_i' \to z_0'$ because $P^+(z_i, x_i) \to \underline{g}^+$ as $i \to \infty$. Thus $z_0' \varepsilon K_\infty(\underset{\to}{r}, y_0')$. It follows immediately that $z_i \varepsilon \underline{g}^+ \cap K_i$ for $i = 1,2$ implies that $z_1 z_2 = y_1 y_2$. One thing remains to be shown.

If $\underline{g}^+$ is a null line, then $\underline{g}^+$ does not intersect $R(\underline{r}^+)$. Let z_i be a sequence of points $\to \infty$ monotonically on $\underline{g}^-$. Then $z_i x_j \to \infty$ as $i \to \infty$ for all j. If y is some point on $\underline{r}^+$ such that $K_\infty(\underset{\to}{r}, y)$ intersects $\underline{g}^+$ at a point z_0', then as before $z_i' = \underline{g}^+ \cap K(x_i, y)$ exist for sufficiently large i, and $z_1' \to z_0'$. Let $y' <_1 y$ be another point on $\underline{r}^+$. Then $K(x_i, y)$ for sufficiently large i intersects $P(x_i, z_i')$ at z_i'', because $P(x_i, z_i')$ eventually are geodesics. Thus $K(x_i, y') \cap \underline{g}^+ = z_i^*$ exist, for such values of i, and $z_i^* <_1 z_{i+1}^* <_1 z_0'$ so that as $i \to \infty$, $z_i^* \to$ some point z_0^* on $\underline{g}^+ \cap K_\infty(\underset{\to}{r}, y')$. Now $P(x_i, z_i') \to \underline{g}^+$, hence $\lim z_i'' = \lim P(x_i, z_i') \cap \lim K(x_i, y') = \underline{g}^+ \cap K_\infty(\underline{r}, y') = z_0^*$. Hence $z_0' z_0^* = \lim z_i' z_i'' = yy' > 0$, a contradiction to the assumption that $\underline{g}^+$ is a null line.

(V.13) The limit circles to a geodesic ray $\underset{\to}{r}$ cover simply an open set $R_\infty(\underset{\to}{r}) = R_\infty(\underline{r}^+)$ such that if p is any point in $R_\infty(\underset{\to}{r})$, then $\underline{A}^+(p, \underset{\to}{r})$ is not a null line, and $\underline{A}^+(p,\underset{\rightarrow}{r}) \subset R_\infty(\underset{\rightarrow}{r})$. Moreover, $R(\underset{\rightarrow}{r}) \subset R_\infty(\underset{\rightarrow}{r}) \subset R'(\underset{\rightarrow}{r})$.

<u>Proof</u>: From (V.11) any asymptote intersecting $R(\underset{\rightarrow}{r})$ intersects some limit circle, and hence by (V.12) it intersects all limit circles, since it is not a null line. Thus $R(\underset{\rightarrow}{r}) \subset R_\infty(\underset{\rightarrow}{r})$. Clearly, $R_\infty(\underset{\rightarrow}{r}) \subset R'(\underset{\rightarrow}{r})$.

(V.14) On each side of a geodesic ray $\underset{\rightarrow}{r}$, the boundary of $R_\infty(\underset{\rightarrow}{r})$ consists of at most two non-intersecting null lines asymptotic to $\underset{\rightarrow}{r}$, one of which, say $\underline{M}^+$, belongs to the boundary of $R'(\underset{\rightarrow}{r})$, and the other belongs to the family of null lines covering R simply, of which $\underline{M}^+$ is not a member.

<u>Proof</u>: Let $\underline{M}^+$ be the null asymptote to $\underset{\rightarrow}{r}$ which is the boundary of $R'(\underset{\rightarrow}{r})$ on one side S of $\underline{r}$. It may not exist, but the following argument is still valid. The null lines intersecting $\underline{M}^+$ cannot be asymptotes to $\underset{\rightarrow}{r}$, hence all points on these null lines and inside $R'(\underline{r})$ lie on geodesic asymptotes to $\underset{\rightarrow}{r}$. Thus $\underline{M}^+$ is the limit of portions of geodesic asymptotes to $\underset{\rightarrow}{r}$ in $R'(\underset{\rightarrow}{r})$ that intersect $R(\underset{\rightarrow}{r})$. Hence $\underline{M}^+$ lies on the boundary of $R_\infty(\underset{\rightarrow}{r})$. Let F_M, F_N be the two families of null lines each covering R simply, with $\underline{M}^+ \in F_M$, and the member of F_M or F_N passing through a point p being denoted by $M(p)$ or $N(p)$ respectively. Then for all points p in S, $M(p)$ intersects $\underline{r}$ and cannot be an asymptote to $\underset{\rightarrow}{r}$.

If $(R'(\underset{\rightarrow}{r}) - R_\infty(\underset{\rightarrow}{r})) \cap S$ is non-empty, then there is a boundary point of $R_\infty(\underset{\rightarrow}{r})$ in $R'(\underset{\rightarrow}{r})$, say p, which is the limit point of a sequence of points $\{p_i\}$ in $R_\infty(\underset{\rightarrow}{r})$ where $\underline{A}^+(p_i,\underset{\rightarrow}{r})$ are geodesics. Hence $\underline{A}^+(p,\underset{\rightarrow}{r})$ must be a null line $N(p)$ contained in $R'(\underset{\rightarrow}{r}) - R_\infty(\underset{\rightarrow}{r})$.

Let ϕ be so chosen that $\underset{\rightarrow}{r}$ is ordered by $<_1$, then a linear ordering $<_1$ can be defined on F_N: $N_1 <_1 N_2$ if there exists p_1 on N_1 and p_2 on N_2 such that $p_1 <_1 p_2$. If some N in F_N is contained in

$R'(\underset{\rightarrow}{r}) - R_\infty(\underset{\rightarrow}{r})$, then so are all $N' <_1 N$ in F_N. Thus there exists such an N such that $N' <_1 N$ for all $N' \varepsilon F_N$ contained in $R'(\underset{\rightarrow}{r}) - R_\infty(\underset{\rightarrow}{r})$. It is then obvious that N lies on the boundary of $R_\infty(\underset{\rightarrow}{r})$ in S, which is therefore $N \cup \underline{M}^+$.

Definition: Let Q be a quadrant of a straight plane R at a point p, and $\bar{Q}$ its closure. Let S be any set in R. A point q in the closure of $S \cap \bar{Q}$ is a foot of p on S if $pq \geq ps$ for all $s \neq q$ on $S \cap \bar{Q}$. We call q a unique foot if $pq > ps$ for all $s \neq q$ in $S \cap \bar{Q}$.

(V.15) If q is a foot of a point $p \neq q$ on a set S, then it is a unique foot of any $r \varepsilon \alpha(p,q) - p$ on S.

Proof: Let s be any point $\neq q$ in S in the same quadrant at r as q. Let ϕ be so chosen that $\alpha^+(q,p)$ is ordered by $<_1$. Then $s <_1 r <_1 p$. If $s \varepsilon \alpha(p,q)$, then $ps < pq$, so that $rs < qr$; if s does not lie on $\alpha(p,q)$, then $rs + pr < ps \leq pq$, so that $rs < pq - pr = qr$. Thus q is a unique foot of r on S.

(V.16) Let $\underset{\rightarrow}{r}$ be a geodesic ray and $\underline{g}^+$ a geodesic asymptote to $\underset{\rightarrow}{r}$ in $R'(\underset{\rightarrow}{r})$. Let $K_\infty(\underset{\rightarrow}{r},y)$ be a limit circle with respect to $\underset{\rightarrow}{r}$ that intersects $\underline{g}^+$ at a point z. Then z is a unique foot of each point in $\underline{g}^+$ on $K_\infty(\underset{\rightarrow}{r},y)$.

Proof: Without loss of generality, assume $\underset{\rightarrow}{r}$ is ordered by $<_1$. Let $K = K_\infty(\underset{\rightarrow}{r},y)$. Let q_i be a sequence of points on $\underset{\rightarrow}{r} \rightarrow \infty$ monotonically along $\underset{\rightarrow}{r}$. Let q be any point on $\underline{g}^+$. We want to prove that z is a foot of q on K, and hence is a unique foot by (V.15).

Case 1. $z <_1 q$. If i is sufficiently large, then $q <_1 q_i$, and eventually $P(q,q_i)$ are geodesics. Thus $P(q_i,q)$ intersects $K(q_i,y)$ at p_i. As $i \to \infty$, $P^+(q,q_i) \to \underline{g}^+$, and $K(q_i,y) \to K$, so that $p_i \to z$. Let x be any point $\neq z$ on K with $x <_1 q$. Then eventually $x <_1 q_i$ and $P(x,q_i)$ are geodesics. Each $P(q_i,x)$ intersects $K(q_i,y)$ at some point $x_i <_1 x$. From $x_i <_1 x <_1 q <_1 q_i$ we have $qx < qx_i \leq q_ix_i - q_iq = p_iq_i - qq_i = p_iq$. We obtain $qx \leq qz$. Thus, z is a foot of q on K.

Case 2. $q <_1 z$. Eventually $z <_1 q_i$ and $P(q,q_i)$ are geodesics. Each $\alpha(q,q_i)$ then intersects $K(q_i,y)$ at p_i which $\to z$ as $i \to \infty$. Let x be any point $\neq z$ on K with $q <_1 x$. There exists an i_0 such that for all $i > i_0$, $x <_1 q_i$ and $P(q_i,x)$ are geodesics, each intersecting $K(q_i,y)$ at some point $x_i <_1 x$ which $\to x$ as $i \to \infty$. Thus for sufficiently large i, we have $q <_1 x_i$, so that $qx_i < qq_i - q_ix_i = qq_i - p_iq_i = p_iq$. As $i \to \infty$, we obtain $qx \leq qz$. Therefore z is a foot of q on K.

(V.17) Let $\underset{\to}{r}$ be a geodesic ray passing through a point x_0, and x_i be a sequence of points $\to \infty$ monotonically on $\underset{\to}{r}$. Let $K = K_\infty(\underset{\to}{r},x_0)$, and $y_0 \neq x_0$ be any point on K. Let a sequence of points y_i converge to y_0, and $K_i = K(x_i,y_i)$. Then $K_i \to K$ as $i \to \infty$.

Proof: Without loss of generality, let us assume $\underset{\to}{r}$ is ordered by $<_1$, $y_0 <_1 x_i$, and $y_i <_1 x_i$. Consequently, by (V.13) $\underline{A}^+(y_0,\underset{\to}{r})$ is a geodesic, so that eventually $P(x_i,y_0)$ are geodesics, intersecting K_i at points which we rename as y_i without loss of generality, and $y_i \to y_0$ as $i \to \infty$ because $\underline{A}^+(y_0,\underset{\to}{r})$ is a geodesic.

Let $K_i' = K(x_i,x_0)$. Then $P(x_i,y_0)$ intersects K_i' at some point z_i, $z_i \to y_0$ as $x_i \to \infty$ on $\underset{\to}{r}$ because $\underline{A}^+(y_0,\underset{\to}{r})$ is a geodesic. Thus $z_iy_i \to 0$, and $x_0x_i' = y_iz_i \to 0$. Here $x_i' = P(x_0,x_i) \cap K_i$.

Let $y \in K$, then eventually $y <_1 x_i$, and $\underline{A}^+(y,\underset{\rightarrow}{r})$ is a geodesic. Let $P(x_i,y)$ intersect K_i and K_i' at y_i' and z_i' respectively. Then $z_i' y_i' = x_0 x_i' = y_i z_i \to 0$ as $i \to \infty$, so that $z_i' \to y$ implies $y_i' \to y$, showing that $K \subset \underline{\lim}\, K_i$.

Suppose $y_i'' \in K_i$ is a sequence of points converging to a point y_0'' in $R_\infty(\underset{\rightarrow}{r})$. Then $\underline{A}^+(y_0'',\underset{\rightarrow}{r})$ is a geodesic, so that eventually $P(x_i,y_i'')$ are geodesics. Thus $P(x_i,y_i'')$ intersects K_i' at z_i'' which $\to$ some point z_0 on K as $i \to \infty$. Hence $z_i'' y_i'' = x_0 x_i' \to 0$, and y_0'' lies on K, showing that $\overline{\lim}\, K_i \cap R_\infty(\underset{\rightarrow}{r}) \subset K$.

Suppose $\overline{\lim}\, K_i - R_\infty(\underset{\rightarrow}{r})$ is non-empty, containing some point c. Since K is a closed set, c has a neighborhood U disjoint from K. By (V.2) K intersects both $S_2(c)$ and $S_4(c)$ because K is ordered by $<_2$. This means that the set $P = K \cap (S_1(c) \cup S_3(c))$ is compact. Each K_i has a portion P_i which $\to K \cap (S_1(c) \cup S_3(c))$, as $i \to \infty$. Since K_i are ordered by $<_2$ it is obvious that c cannot exist. Thus $\overline{\lim}\, K_i \subset K$. Combining with the results of the former paragraphs, we have $\lim K_i = K$.

This lemma enables us to define a limit circle as $K_\infty(\underset{\rightarrow}{r},y) = \lim_{i\to\infty} K(x_i,y_i)$ where $x_i \to \infty$ on $\underset{\rightarrow}{r}$ and $y_i \to$ some y in $R_\infty(\underset{\rightarrow}{r})$, not necessarily lying on the geodesic $\underline{r}^+$ which contains $\underset{\rightarrow}{r}$.

<u>Definition</u>: Let an oriented geodesic $\underline{r}^+$ be represented by $z(t)$ with its arc-length t as parameter, so that $z(t_1)z(t_2) = |t_1 - t_2|$ is the metric distance between the two points $z(t_1)$ and $z(t_2)$ on $\underline{r}^+$. Let $\underset{\rightarrow}{r}$ be any ray on $\underline{r}^+$ with the same orientation, and x be any point in R. Then $\gamma(\underline{r}^+,x) = \gamma(\underset{\rightarrow}{r},x)$ is defined as $\lim_{t\to\infty} (xz(t) - t)$, if the limit exists.

(V.18) If x is any point in $R_\infty(\underline{r}^+)$, then $K_\infty(\underline{r}^+,x) = \{y \mid \gamma(\underline{r}^+,y) =$

$\gamma(\underline{r}^+,x)\} \cap R_\infty(\underline{r}^+)$.

Proof: Let $\underline{r}^+$ be reparametrized by $z(t)$ so that $z(0) = K_\infty(\underline{r}^+,x) \cap \underline{r}^+$. We shall prove that $\gamma(\underline{r}^+,p) = 0$ for any point p on $K_\infty(\underline{r}^+,z(0))$. Now $\gamma(\underline{r}^+,p) = \lim_{t\to\infty} (pz(t) - t) = \lim_{t\to\infty} (pz(t) - z(0)z(t))$. As $t \to \infty$, $P(p,z(t))$ eventually are geodesics, hence intersect $K(z(t), z(0))$ at a point $p(t)$ such that $\lim p(t) = p$. Thus $\lim_{t\to\infty} (pz(t) - z(0)z(t)) = \lim_{t\to\infty} (pz(t) - p(t)z(t))$ $= \lim (-pp(t)) = 0$.

Since $R_\infty(\underline{r}^+)$ is covered simply by limit circles, only the points p on $K_\infty(\underline{r}^+,z(0))$ satisfy $\gamma(\underline{r}^+,p) = 0$.

(V.19) If $\underline{r}^+$ is an oriented geodesic ordered by $<_1$ and $x <_1 y$ are two points in $R_\infty(\underline{r}^+)$, then $\gamma(\underline{r}^+,y) - \gamma(\underline{r}^+,x) \geq xy$, equality holding only when $y \in \underline{A}^+(x,\underline{r}^+)$.

Proof: This follows from (V.16).

(V.20) If $\underline{r}^+$ is an oriented geodesic, then another oriented geodesic $\underline{g}^+$ parametrized by its arclength as $w(t)$ is an asymptote in $R_\infty(\underline{r}^+)$ to $\underline{r}^+$ if for some $t_1 > t_2$, $w(t_1)$ and $w(t_2)$ are points in $R_\infty(\underline{r}^+)$ and that $t_1 - t_2 = \gamma(\underline{r}^+,w(t_1)) - \gamma(\underline{r}^+,w(t_2))$.

4. Parallels in a Straight Plane. Throughout this section, let R be a straight plane, in which the following parallel axiom is true:

If $\underline{g}^+$ is an asymptote to an oriented path $\underline{h}^+$, then $\underline{g}^-$ is an asymptote to $\underline{h}^-$. (Both $\underline{g}^+$ and $\underline{h}^+$ may be null lines.)

$\underline{g}^+$ is called a parallel to $\underline{h}^+$, and is said to be parallel to $\underline{h}^+$, denoted by $\underline{g}^+ \| \underline{h}^+$. The orientation of $\underline{g}^+$ is said to be parallel to $\underline{h}^+$, and the orientation of $\underline{g}^-$ is said to be reversed to $\underline{h}^+$.

(V.21) Theorem. Let R be a straight plane in which the parallel axiom holds. Then:

1.) The parallel relation among oriented paths is symmetric and transitive.

2.) Parallels to null paths are null paths, and parallels to geodesics are geodesics. Every null line intersects all geodesics, and every geodesic intersects all null lines.

3.) A grid-representation ϕ of R maps R into the interior of some euclidean grid-rectangle S lying in the interior of the square with the points $(\pm 1, \pm 1), (\pm 1, \mp 1)$ as vertices.

4.) If $\underline{g}$ is a geodesic, then $\phi\underline{g}$ has a pair of opposite vertices of S as endpoints.

Proof: If $\underline{g}^+$ is an oriented path, and p any point not lying on $\underline{g}^+$, then there exists only one path through p that does not intersect $\underline{g}^+$, namely, the parallel to $\underline{g}^+$ through p. Thus a necessary and sufficient condition for an oriented path to be parallel to another is that they do not intersect. It follows immediately that parallel relation among oriented paths is symmetric. To see that it is transitive, suppose $\underline{g}^+ \| \underline{h}^+$, $\underline{h}^+ \| \underline{k}^+$ are three oriented paths. The $\underline{g}^+$ does not intersect $\underline{k}^+$, otherwise if p is a point lying on both $\underline{g}^+$ and $\underline{k}^+$, then $\underline{g}^+$ and $\underline{k}^+$ are two different parallels to $\underline{h}^+$ through p, contradicting the parallel axiom. Thus $\underline{g}^+$ does not intersect $\underline{k}^+$, and so $\underline{g}^+ \| \underline{k}^+$.

It follows that parallels to null lines must be null lines of the same simple family, and geodesics are never parallels to null lines, hence every null line intersects all geodesics. Similarly, a null line can never be a parallel to a geodesic, so that each geodesic intersects all null lines.

Let $\underline{g}^+$ be an oriented geodesic. We assert $R'(\underline{g}^+) = R$. For if not, then the boundary of $R'(\underline{g}^+)$ consists of one or more null lines not intersecting $\underline{g}^+$, contradicting the last paragraph. Thus $R'(\underline{g}^+) = R$, and similarly $R'(\underline{g}^-) = R$, so that $R(\underline{g}^+) = R'(\underline{g}^+) \cap R'(\underline{g}^-) = R$. Let ϕ be any grid-representation of R such that $\underline{g}^+$ is ordered by $<_1$. Let $\underline{h}^+$ be any geodesic ordered by $<_2$. Let the endpoints of $\phi\underline{g}^+$ be $(x_1,y_1),(x_3,y_3)$ in E^2 where $x_1 < x_3$, $y_1 < y_3$, and the endpoints of $\phi\underline{h}^+$ be $(x_4,y_4),(x_2,y_2)$ where $x_4 < x_2$, $y_4 > y_2$. Then ϕR is contained and circumscribed by the grid-rectangle with $(x_1,y_1),(x_3,y_3)$ as opposite vertices as well as by the grid-rectangle with $(x_4,y_4),(x_2,y_2)$ as vertices, so that $x_1 = x_4$, $x_2 = x_3$, $y_1 = y_2$, $y_3 = y_4$, and ϕR is a rectangle with vertices at $(x_1,y_1),(x_3,y_1)$, $(x_3,y_3),(x_1,y_3)$. Let $\underline{g}$ be any arbitrary geodesic. Then it intersects all null lines, therefore $\phi\underline{g}$ has a pair of opposite vertices of ϕR as endpoints.

The parallel axiom thus has strong implications. We now study its relation to perpendicularity. We write $\underline{g} \dashv \underline{h}$ or $\underline{h} \vdash \underline{g}$ if $\underline{g}$ is perpendicular to a geodesic $\underline{h}$.

(V.22) <u>Theorem</u>. Let R be a straight plane satisfying the parallel axiom in which all circles are concave. Let $\underline{h}_1 \| \underline{h}_2$ be two parallel geodesics. If $\underline{g}$ is a geodesic such that $\underline{h}_1 \dashv \underline{g}$, then $\underline{h}_2 \dashv \underline{g}$ also.

<u>Proof</u>: Let $p_1 = \underline{g} \cap \underline{h}_1$. $\underline{g}$ is not parallel to $\underline{h}_2$, otherwise $\underline{h}_1 \| \underline{h}_2$ and $\underline{g} \| \underline{h}_2$ implies $\underline{h}_1 \| \underline{g}$ which is impossible. Thus $p_2 = \underline{g} \cap \underline{h}_2$ exists. Suppose $\underline{h}_2$ is not perpendicular to $\underline{g}$. Let $q \neq p_2$ be any point on $\underline{h}_2$. Then p_2 is not the foot of q on $\underline{g}$. Since it is trivial that q is in $R(\underline{g})$, q does have a foot f on $\underline{g}$ different from p_2, and depending on q con-

tinuously, so that if q is close to p_2, f is then different from p_1. However, $P(f,q)$ is not parallel to $\underline{h}_1$, hence $r = P(f,q) \cap \underline{h}_1$ exists. Then f and p_1 are both feet of r on $\underline{g}$, contradicting the hypothesis that all circles are concave.

(V.23) In a straight plane R where all circles are concave and the parallel axiom is true, let $\underline{h}_1$ be a perpendicular to a geodesic $\underline{g}$ intersecting $\underline{h}_1$ at p. Let f be any point on $\underline{g}$, then there exists a unique perpendicular line $\underline{h}_2$ to $\underline{g}$ through f, namely, the parallel to $\underline{h}_1$ through f. If $\underline{h}_3 \neq \underline{h}_2$ is any other path intersecting $\underline{g}$ at f, and q is any point on $\underline{g}$, then there exists uniquely a point r on $\underline{h}_3$ such that q is the foot of r on $\underline{g}$.

<u>Proof</u>: The parallel $\underline{h}_2$ to $\underline{h}_1$ through f being a perpendicular to $\underline{g}$, any other path $\underline{h}_3$ intersecting $\underline{g}$ at f is not a parallel to $\underline{h}_1$, hence $s = \underline{h}_1 \cap \underline{h}_3$ exists. If $\underline{h}_3$ were a perpendicular to $\underline{g}$, then f would be a foot of s on $\underline{g}$, so that f and p are both feet of s on g, which is impossible. The parallel $\underline{h}_4$ through q to $\underline{h}_1$ must intersect $\underline{h}_3$ at some point r, otherwise $\underline{h}_4 \| \underline{h}_3$, so that $\underline{h}_1 \| \underline{h}_3$, which is impossible. Hence $\underline{h}_4 \dashv \underline{g}$, and q is the foot of r on $\underline{g}$.

(V.24) Let a,b be two points on a limit circle $K = K_\infty(\underline{g}^+,q)$ through a point q on a geodesic $\underline{g}^+$. Let $\underline{h}^+$ be the parallel to $\underline{g}^+$ through a. Then $\underline{h}^+$ is perpendicular to the geodesic $P(a,b)$.

<u>Proof</u>: If $\underline{h}^+$ is not perpendicular to $P(a,b)$, then from (V.23) there exists a point c on $\underline{h}^+$ with b as its foot on $P(a,b)$. Then $bc > ac$, contradicting the fact that a is the foot of any point of $\underline{h}^+$ on K.

(V.25) <u>Theorem</u>. If a straight plane R satisfies the parallel axiom and has concave circles, then a geodesic $\underline{g}$ which is perpendicular to one of two parallel geodesics $\underline{h}_1 \| \underline{h}_2$ is also perpendicular to the other.

Proof: (Similar to the proof of Theorem 24.2 in Busemann [3].) Assume ϕ is so chosen that $\underline{g}$ is ordered by $<_1$ and $\underline{h}_1$ and $\underline{h}_2$ are ordered by $<_2$. Assume $\underline{g} \dashv \underline{h}_1$. Let $p_1 = \underline{h}_1 \cap \underline{g}$. Then $\underline{g}$ is not parallel to $\underline{h}_2$, hence $p_2 = \underline{h}_2 \cap \underline{g}$ exists. If one orientation $\underline{g}^+$ of $\underline{g}$ is perpendicular to $\underline{h}_1$, then so is $\underline{g}^-$. Thus we may assume without loss of generality that $p_2 <_1 p_1$.

Suppose it is not true that $\underline{g} \dashv \underline{h}_2$. Then there exists a geodesic $\underline{k}$ through p_2 ordered by $<_2$ such that $p_2 = \underline{k} \cap K(p_1, p_2)$, so that p_2 is the foot of p_1 on $\underline{k}$ and $\underline{g} \dashv \underline{k}$. $\underline{k}$ intersects $\underline{h}_1$ at some point r. Assume ϕ is so chosen that $p_2 <_2 r$. Let $r <_2 q$ be any point on $\underline{h}_1$. Let $\underline{k}' = P(p_2, q)$. We want to prove that $\underline{g} \dashv \underline{k}'$, so that as $p_1 q \to \infty$, $\underline{k}' \to \underline{h}_2$, so that p_2 is the foot of p_1 on $\underline{h}_2$, and $\underline{g} \dashv \underline{h}_2$ would be proved.

We shall prove that any point on $\underline{k}'$ different from p_2 is not a foot of p_1 on $\underline{k}'$. Let $x <_1 p_1$ be a point on $\underline{k}'$ such that $x <_2 p_2$. Then $P(p_1, x) \cap \underline{k} = x_0$ exists, so that $p_1 x < p_1 x_0 < p_1 p_2$. Thus x is not a foot of p_1 on $\underline{k}'$.

Let y be any point on $\underline{k}'$ such that $p_2 <_2 y <_1 p_1$. Let $u <_2 q$ be any point on $\underline{h}_1$ different from q. The parallel $\underline{g}'$ through u to $\underline{g}$ and the parallel through q to $P(p_1, y)$ are not parallel, hence they intersect at some point w with $u <_1 w$, $q <_1 w$. $\underline{g}'$ also intersects $\underline{k}'$ at a point v with $v <_1 u$. By the last theorem $\underline{g}' \dashv \underline{h}_1$, hence $vw > uw > qw$. Thus q is not a foot of w on $\underline{k}'$, and it is not true that $P(q,w) \dashv \underline{k}'$. Hence $P(p_1, y) \dashv \underline{k}'$ is not true. It is trivial that $p_1 \in R(\underline{k}') = R$. Hence p_1 must have a foot on $\underline{k}'$, namely, p_2. This proves the theorem.

Definition: A straight space R is said to Desarguesian if there exists a homeomorphism $\phi: R \to E^2$ such that all paths are mapped into euclidean lines. We call ϕ a Desarguesian map.

(V.26) If R satisfies the parallel axiom, then $\phi R = E^2$ for any Desarguesian map.

Proof: Any two points in ϕR can be joined by a euclidean segment $\subset \phi R$. Therefore ϕR is a convex region in E^2. Suppose $\phi R \neq E^2$, then the boundary of ϕR is non-empty. Let two euclidean lines intersect at a point p on the boundary of ϕR. Then in ϕR they do not intersect, so they must be parallel, but then it is obvious that there exist other euclidean lines intersecting the first line in ϕR but do not intersect the second in ϕR. Then the parallel axiom is contradicted.

Definition: Let K be a component of a circle K'. A geodesic $\underline{g}$ passing through a point p on K such that K lies on one side of $\underline{g}$ is called a supporting line of K and of K'. K' and K are differentiable at p if the supporting line through p is unique.

(V.27) In a straight plane R satisfying the parallel axiom, where all circles are concave, if $P^+(p,q)^+ = \underline{g}^+$ is a geodesic, and $K(p,q)$ is differentiable at q, then $K_\infty(\underline{g}^-,q)$ is a straight line.

Proof: Without loss of generality assume $\underline{g}^+$ is ordered by $<_1$. Let r be a variable point $\to \infty$ along $\underline{g}^-$. Then $K_\infty(\underline{g}^-,q) = \lim_{r\to\infty} K(r,q)$. For any $r <_1 q$, $K(r,q)$ has a unique supporting line $\underline{k}(r)$ through q. If $r_1 <_1 r_2 <_1 q$, and $\underline{k}(r_1) \neq \underline{k}(r_2)$, then $\underline{k}(r_1)$ is a supporting line of $K(r_2,q)$, contradicting the differentiability of $K(r_2,q)$ at q. Hence $\underline{k}(r)$

is independent of r on $\underline{g}^-$. As $r \to \infty$ along $\underline{g}^-$, $\underline{k} = \underline{k}(r)$ becomes a supporting line of $K_\infty(\underline{g}^-,q)$. Let s be any point on $K_\infty(\underline{g}^-,q)$, then q is the foot of p on $P(q,s)$, so that $P(q,s)$ is a supporting line of $K(p,q)$. Thus $P(q,s) = \underline{k}$ so that $K_\infty(\underline{g}^-,q) = \underline{k}$.

(V.28) For a geodesic $\underline{g}^+$, let $\underline{h}$ be a supporting line of $K(p,q)$ for some p,q on $\underline{g}^+$, at the point q. Then for all r,s on $\underline{g}$, if $\underline{h}'$ is the parallel to $\underline{h}$ through s, then $\underline{h}'$ is a supporting line of $K(r,s)$, $K_\infty(\underline{g}^+,s)$ and $K_\infty(\underline{g}^-,s)$, for all s.

Proof: q is a foot of p on $\underline{h}$, so that $\underline{g}^+ \dashv \underline{h}$. Hence q is the foot of any point r on $\underline{g}^+$, so that $\underline{h}$ is a supporting line of $K(r,q)$ for all r on $\underline{g}^+$. As $r \to \infty$ along $\underline{g}^+$ or $\underline{g}^-$, $\underline{h}$ becomes a supporting line for $K_\infty(\underline{g}^+,q)$ and $K_\infty(\underline{g}^-,q)$. Since $\underline{h}' \| \underline{h}$, s is a foot of p and hence of any point r of $\underline{g}^+$ on $\underline{h}'$. Thus $\underline{h}'$ is a supporting line for $K(r,s)$ for all r,s on $\underline{g}^+$.

(V.29) Theorem. If a straight plane R satisfies the parallel axiom, all circles are concave, and R is Desarguesian, then R is Minkowskian.

Proof: (This is different from the proof of the analogous theorem 24.1 of Busemann [3], because we do not have an analogue to theorem 16.7, ibidem, in doubly-timelike spaces.)

Let R be represented as the whole euclidean (x,y)-plane in which all paths are euclidean lines. We shall not distinguish between a subset of R and its image in E^2 under the Desarguesian map. This proof is divided into several parts:

1.) Let p be any point in R. We assert that all circles with p as center can be obtained one from the other by euclidean dilations keeping

p as well as all paths (i.e., euclidean lines) through p invariant.

Without loss of generality, let $K_1 = K(p,q_1)$, $K_2 = K(p,q_2)$ be the components of two circles in $S_1(p)$, and assume p is the origin. Let $\underline{g}$ be any ray in $S_1(p)$ initiating from p and Θ the angle it makes with the positive x-axis. Then $\underline{g}$ depends on Θ. Let $\underline{g}$ intersect K_1 and K_2 at $r_1(\Theta)$ and $r_2(\Theta)$. Let ρ_i be the euclidean distance between p and r_i for $i = 1,2$. We want to prove that $\rho_2(\Theta)/\rho_1(\Theta)$ is independent of Θ. $\rho_i(\Theta)$ are continuous and never zero for $0 < \Theta < \pi/2$. Since circles in R are concave, $\rho_i(\Theta)$ are non-differentiable at only a countable number of points, where the left- and right-derivatives $D_-\rho_i(\Theta)$ and $D_+\rho_i(\Theta)$ still exist, with $D_-\rho_i(\Theta) < D_+\rho_i(\Theta)$, the curve $\rho_i(\Theta)$ in the (ρ,Θ)-plane being concave upwards for each i.

From (V.28) we see that for each Θ, $D_-\rho_1(\Theta) = D_-\rho_2(\Theta)$ and $D_+\rho_1(\Theta) = D_+\rho_2(\Theta)$, so that either both $\rho_1(\Theta)$ and $\rho_2(\Theta)$ are differentiable, when $\rho_1'(\Theta)/\rho_1(\Theta) = \rho_2'(\Theta)/\rho_2(\Theta)$, or both are non-differentiable, when $D_-\rho_1(\Theta)/\rho_1(\Theta) = D_-\rho_2(\Theta)/\rho_2(\Theta)$ and $D_+\rho_1(\Theta)/\rho_1(\Theta) = D_+\rho_2(\Theta)/\rho_2(\Theta)$. Let $f_i(\Theta) = \ln \rho_i(\Theta)$, then ρ_i are differentiable if, and only if, $f_i(\Theta)$ are so, and $D_-f_1(\Theta) = D_-f_2(\Theta) \leq D_+f_1(\Theta) = D_+f_2(\Theta)$, equality holding only when $f_i(\Theta)$ are differentiable. It follows that $f_1(\Theta) = f_2(\Theta) + \text{constant}$, and so $\rho_1(\Theta) = k\rho_2(\Theta)$ for some positive constant k, as to be proved. The case where K_1 and K_2 lie in opposite quadrants at p is similar.

Let $\underline{g}$ be any geodesic, then we have proved that either $K(p,q)$ is differentiable at q for all q,p on $\underline{g}$ or for no pair of points p,q on $\underline{g}$ at all. We call $\underline{g}$ <u>ordinary</u> or <u>exceptional</u> respectively.

2.) Let $\underline{g}^+$ be any oriented geodesic ordered by $<_1$. Let q_1,q_2 be two fixed points on $\underline{g}^+$, and p a variable point $\to \infty$ along $\underline{g}^-$. For any position of p, there exists a euclidean dilation keeping p and all paths through p invariant, and mapping $K(p,q_1)$ onto $K(p,q_2)$. As $p \to \infty$, the paths through p become parallel geodesics to $\underline{g}^-$, and $K(p,q_i) \to K_\infty(\underline{g}^-,q_i)$ for $i = 1,2$, so that there exists a euclidean translation of R along $\underline{g}^-$ mapping q_1 to q_2, $K_\infty(\underline{g}^-,q_1)$ onto $K_\infty(\underline{g}^-,q_2)$, and keeping all the parallel paths to $\underline{g}^-$ invariant.

3.) We now show that on a geodesic $\underline{g}^+$, the metric distance is proportional to the euclidean distance. If q,q' are any two points on $\underline{g}^+$, their metric distance and euclidean distance are denoted by qq' and $e(q,q')$ respectively. Let p be a point on $\underline{g}^+$, which is assumed ordered by $<_1$ without loss of generality. Let $\mu = pq$, $s = e(p,q)$ for all points q on $\underline{g}^+$ with $p <_1 q$, and $\mu = -pq$, $s = -e(p,q)$ if $q <_1 p$. Then μ is a monotonically strictly increasing continuous function $\mu(s)$ of s such that $\mu(0) = 0$. We want to prove that $\mu = \gamma s$ for some γ independent of s. We shall say s and μ are the signed euclidean distance and the signed metric distance on $\underline{g}^+$ with respect to p.

Let $p <_1 q$ be any point on $\underline{g}^+$. Let $\underline{g_2}^+$ be another geodesic through p ordered by $<_1$ and intersecting $K(p,q)$ at $q <_2 q_2$ such that $\underline{g_2}^+$ is ordinary. By (V.6) $K(p,q)$ is strictly convex, and the supporting line $\underline{k}$ of $K(p,q)$ at q_2 is not a supporting line of $K(p,q)$ at q. Let $\underline{h} = K_\infty(\underline{g}^+,q)$, and $\underline{k}'$ be a geodesic through p parallel to $\underline{k}$ and $\underline{h}' = K_\infty(\underline{g}^+,p)$. Let p_3 be any point with $p <_2 p_3$ on h'. Let $\underline{g_3}^+$ and $\underline{g_4}^+$ be oriented geodesics through p_3 parallel to $\underline{g}^+$ and $\underline{g_2}^+$ respectively. Let $q_3 = \underline{h} \cap \underline{g_3}^+$, $q_4 = \underline{k} \cap \underline{g_4}^+$, $p_4 = \underline{k}' \cap \underline{g_4}^+$, $p_2 = p$.

Let μ_i be the signed metric distance on $\underline{g_i}^+$ with respect to p_i for $i = 2,3,4$, and μ_i is a monotonically strictly increasing continuous function of the signed euclidean distance on $\underline{g}_i^+$ with respect to p_i.

From part 2.) of this proof, we see that $\mu_3(s) = \mu(s)$ and $\mu_4(s) = \mu_2(s - s_0) + \mu_4(s_0)$ where $s_0 = e(p_3, p_4)$.

From part 1.) of this proof, we have $\mu_2(s) = \mu(\gamma s)$, $\mu_4(s) = \mu_3(\eta s)$ for some constants γ, η independent of s. Thus $\mu(\eta s) = \mu_3(\eta s) = \mu_4(s) = \mu_2(s - s_0) + \mu_4(s_0) = \mu(\gamma s - \gamma s_0) + \mu_3(\eta s_0) = \mu(\gamma s - \gamma s_0) + \mu(\eta s_0)$. Let $\sigma = \eta s$, $\sigma_0 = \eta s_0$, then $\mu(\sigma) = \mu(\frac{\gamma}{\eta}(\sigma - \sigma_0)) + \mu(\sigma_0)$, which will be referred to as (*).

It is easy to see that $\gamma/\eta = 1$, for if not, then we can choose $\sigma = \sigma_0\gamma/(\gamma - \eta)$, so that $(\gamma/\eta)(\sigma - \sigma_0) = \sigma$, and (*) becomes $\mu(\sigma) - \mu(\sigma_0) = \mu(\sigma)$, or $\mu(\sigma_0) = 0$, which is impossible.

Let $\nu(\sigma) = \mu(\sigma) - \sigma\mu(\sigma_0)/\sigma_0$, then $\nu(0) = \nu(\sigma_0) = 0$, and (*) becomes on simplification $\nu(\sigma) = \nu(\sigma_0) + \nu(\sigma - \sigma_0)$ or $\nu(\sigma) = \nu(\sigma - \sigma_0)$, so that ν is a periodic function with period σ_0. Now let p_3 move along h' to p. $\sigma_0 = \eta s_0$ will vary continuously to zero, but $\nu(\sigma) = \nu(\sigma - \sigma_0)$ remains true for all p_3. This means $\nu(\sigma) = 0$ for all σ, i.e., $\mu(\sigma) = \sigma\mu(\sigma_0)/\sigma_0$, hence $\mu(s)$ is proportional to s. R is therefore a Minkowski plane with strictly concave circles.

CHAPTER VI

FLAT BISECTORS

1. Bisectors with Flat Subsets. Given two distinct points a_1, a_2 in a G-space, the bisector of a_1 and a_2 is $B(a_1,a_2) = \{x \mid a_1x = a_2x\}$. A bisector B in a G-space is called flat if $x,y \in B$ implies a segment from x to y also lies in B. Consideration of the two dimensional Lorentz space shows that to require $B(a_1,a_2)$ to be flat according to this definition is far too strong a condition for general G-spaces.

In this chapter the space R will always be a doubly timelike surface. The symbol U_0 will denote a null box in R with $\bar{U}_0 \times \bar{U}_0 \subset \Pi_0$. $P(a_1,a_2)$ denotes the "line" in U_0 through a_1 and a_2. The bisector in U_0 of a_1 and a_2 will be $B(a_1,a_2) = \{x \in U_0 \mid a_1x = a_2x\}$.

Definition: Given distinct points $a_1, a_2 \in U_0$ the bisector $B(a_1,a_2)$ is said to have a (proper) locally flat subset S in U_0, provided

1.) S is a line in U_0

2.) $S \cap \alpha(a_1,a_2)$ is a single point q.

Bisector Condition:

A doubly timelike surface R is said to satisfy the bisector condition provided each point $p \in R$ has a neighborhood U_0 such that $a_1, a_2 \in U_0$ with $a_1a_2 > 0$ implies $B(a_1,a_2)$ has a (proper) locally flat subset S.

The null lines of U_0 are divided into two families F_M and F_N that each cover U_0 simply. If $x \in U_0$, then $M(x)$ and $N(x)$ represent the null lines through x that are in F_M and F_N respectively. Given a_1, $a_2 \in U_0$ with $a_1a_2 > 0$ let $a_3 = M(a_1) \cap N(a_2)$ and $a_4 = M(a_2) \cap N(a_1)$. Since U_0 is a null box, the points a_3 and a_4 exist and are in U_0. As before, $Q(a_1,a_2)$ denotes the null box bounded by $\alpha(a_1,a_3)$, $\alpha(a_3,a_2)$, $\alpha(a_2,a_4)$ and $\alpha(a_4,a_1)$.

(VI.1) Let R satisfy the bisector condition and $a_1,a_2 \in U_0$ with $a_1a_2 > 0$. Then $S = P(a_3,a_4)$ and S is not a null line.

<u>Proof</u>: The point $q = S \cap \alpha(a_1,a_2)$ must lie in $Q(a_1,a_2)$. Therefore, S must intersect the boundary of $Q(a_1,a_2)$ in two points. However, $B(a_1,a_2)$ intersects the boundary of $Q(a_1,a_2)$ only in the points a_3 and a_4 so that $S = P(a_3,a_4)$.

The point a_3 has only two null lines $M(a_3)$ and $N(a_3)$ through it. Since S cannot be either of these lines, S cannot be a null line.

(VI.2) Let R satisfy the bisector condition and $a_1,a_2 \in U_0$ with $a_1a_2 > 0$. Then $B(a_1,a_2) \cap \bar{Q}(a_1,a_2) = \alpha(a_3,a_4)$.

<u>Proof</u>: Assume $y' \in B(a_1,a_2) \cap \bar{Q}(a_1,a_2) - \alpha(a_3,a_4)$. If y' is on the same side of $\alpha(a_3,a_4)$ as a_1, let $\alpha(a_2,y') \cap \alpha(a_3,a_4) = y$. For one of the partial orderings $<_i$ in U_0, let $a_1 <_i y' <_i y <_i a_2$. Then

$$a_1y \geq a_1y' + y'y > a_1y' = a_2y' > a_1y = a_2y.$$

Therefore, no such y' can exist. Clearly $\alpha(a_3,a_4) \subset S \cap \bar{Q}(a_1,a_2)$ so the result follows.

According to the definition of perpendicularity given in Chapter II, the segment $\alpha(a_1,a_2)$ will be perpendicular to $\alpha(a_3,a_4)$ at $q = \alpha(a_1,a_2) \cap \alpha(a_3,a_4)$ if for each x in $\alpha(a_1,a_2) - q$, the function $f_x(y) = xy$ for $y \in \alpha(a_3,a_4)$ has a local maximum at $y = q$.

(VI.3) Let R satisfy the bisector condition and $a_1,a_2 \in U_0$ with $a_1a_2 > 0$.

Then $\alpha(a_1,a_2)$ and $\alpha(a_3,a_4)$ are mutually perpendicular at q.

Proof: It is only necessary to show $\alpha(a_1,a_2) \dashv \alpha(a_3,a_4)$. Let $x \in \alpha(a_1,a_2) - q$ and set $u_1 = M(x) \cap \alpha(a_3,a_4)$, $x' = N(u_1) \cap \alpha(a_1,a_2)$ and $u_2 = M(x') \cap N(x)$. By (VI.1) then $\alpha(a_1,a_2) \subset B(u_1,u_2)$ so that $a_1u_1 = a_1u_2$ and $a_2u_1 = a_2u_2$. But $a_1u_1 = a_2u_1$ so $a_1u_1 = a_2u_2$ and by (VI.2) we have $u_2 \in \alpha(a_3,a_4)$. Thus, $P(a_3,a_4) \subset B(x,x')$.

If $y \in \alpha(u_1,u_2) - u_1 - u_2$ in some partial ordering $<_i$ we have $x <_i y <_i x'$. Hence $2xq \geq xy + yx' = 2yx$ and equality only for $y = q$.

Given $\alpha(a_3,q)$ an arclength parametrization $y(s)$ is such that $y(0) = a_3$ and $y(a_3q) = q$, as well as $y(s_1)y(s_2) = |s_1 - s_2|$.

(VI.4) $f(s) = a_1y(s)$ is strictly increasing for $0 \leq s \leq a_1q$.

Proof: Assume $f(s_1) = f(s_2)$ for $s_1 < s_2$. Choose r_1 and r_2 distinct points of $P(a_3,a_4)$ with $r_1y(s_2) = r_2y(s_2)$. Let P be the proper locally flat subset of $B(r_1,r_2)$. Then $P \cap \alpha(a_i\ , y(s_1))$ consists of a point u_i for $i = 1,2$. In some partial ordering $<_j$ in U_0 then $u_1 <_j y(s_1) <_j u_2$ and $u_1 <_j y(s_2) <_j u_2$. But $P(a_3,a_4) \cap U_0 \subset B(u_1,u_2)$ and $\alpha(u_1,u_2) \cap P(a_3,a_4) = y(s_2)$. Therefore, $u_1y(s_2) = u_2y(s_2)$ and also $u_1y(s_2) = u_1y(s_1) = u_2y(s_1)$. This is impossible unless $y(s_1) \in \alpha(u_1,u_2)$, which implies $y(s_1) = y(s_2)$.

Assume R satisfies the bisector condition and let P be a line in U_0 that is not null. At each point $y \in P$ there is exactly one line P_y such that if α and α' are Arcs on P and P_y respectively with y interior to both α and α', then $\alpha \perp \alpha'$. We call P and P_y perpendicular at y and write $P \perp P_y$. [Compare with definition p. 67.]

For a pair $a_1, a_2 \in U_0$ with $a_1a_2 > 0$ let P_y be the perpendicular to $P(a_1,a_2)$ at $y \in \alpha(a_1,a_2)$. Then for each $x \in \bar{Q}(a_1,a_2)$ there is a unique $y \in \alpha(a_1,a_2)$ with $x \in P_y$. If $xy = 0$ then $x = y$ and we set $x' = x = y$. If $xy > 0$ there is a unique point $x' \in P_y \cap \bar{Q}(a_1,a_2)$ with $x \neq x'$ and $xy = x'y$. Define the map $R_{a_1a_2}$, of $\bar{Q}(a_1,a_2)$ onto inself by $R_{a_1a_2}(x) = x'$. From the definition of $R_{a_1a_2}$ it follows that $R^2_{a_1a_2}$ is the identity.

(VI.5) The map $R_{a_1a_2}$ is topological.

<u>Proof</u>: Since it must be one to one and $R^2_{a_1a_2}$ is the identity, it is only necessary to show $R_{a_1a_2}$ is continuous. Let x_ν and $x \in \bar{Q}(a_1,a_2)$ with $x_\nu \to x$. If $x_\nu \in P_{y_\nu}$ and $x \in P_y$ then $y_\nu \to y$ because of uniqueness of feet and continuity of the metric, it then follows that $x_\nu' \to x'$ since $x_\nu y_\nu \to xy$.

(VI.6) Let β be an Arc in $\bar{Q}(a_1,a_2)$ with $\beta \cap \alpha(a_1,a_2) \neq \phi$. Then β is mapped to an Arc β' under $R_{a_1a_2}$. If $x_1, x_2 \in \beta$ then $x_1'x_2' = x_1x_2$.

<u>Proof</u>: Let $r = \beta \cap \alpha(a_1,a_2)$. By continuity r may be taken as a relative interior point of both β and $\alpha(a_1,a_2)$.

Assume (x_1rx_2) so that x_1 and x_2, as well as x_1', lie on opposite sides of $\alpha(a_1,a_2)$. If $u <_i v <_i w$ then $u' <_j v' <_j w'$. Consequently, for some i and j we have $x_1 <_i r <_i x_2$ and $x_1' <_j r' = r <_j x_2'$.

Let $\alpha(x_1',x_2') \cap \alpha(a_1,a_2) = s$, then $x_1' <_j s <_j x_2'$. Since $\alpha(a_1,a_2) \subset B(x_1,x_1') \cap B(x_2,x_2')$, we have $x_1s = sx_1'$ and $x_2s = x_2's$. Also, $x_1r = x_1'r$. Therefore,

$$x_1x_2 = x_1r + rx_2 \geq x_1s + sx_2 = x_1's + sx_2' \geq x_1'r + rx_2' = x_1r + rx_2 = x_1x_2.$$

Equality implies $r = s$ and $x_1'x_2' = x_1x_2$.

If x_1 and x_2 are on the same side of $\alpha(a_1,a_2)$, choose $w \in \beta$ with x_1 and w on opposite sides of $\alpha(a_1,a_2)$. Let $\alpha(x_1,w) \cap \alpha(a_1,a_2) = r$. Then (x_1rw), (x_2rw), $(x_1'rw')$ and $(x_2'rw')$ all hold. Since $x_1'w' = x_1w$ and $x_2'w' = x_2w$, it follows that $x_1'x_2' = x_1x_2$.

Let a_1,a_2,a_3,a_4 and $q = \alpha(a_1,a_2) \cap \alpha(a_3,a_4)$ be as above. Define $R_q = R_{a_3a_4} \circ R_{a_1a_2}$. This is a topological map of $\bar{Q}(a_1,a_2)$ onto itself. Any Arc β in $\bar{Q}(a_1,a_2)$ that intersects both $\alpha(a_1,a_2)$ and $\alpha(a_3,a_4)$ is mapped to an Arc β'. If $x_1,x_2 \in \beta$, then $x_1x_2 = R_q(x_1)R_q(x_2)$.

(VI.7) Let $\beta = \alpha(c_1,c_2)$, $q \in \beta$ and let c_1 and c_2 lie on the boundary of $\alpha(a_1,a_2)$. Then R_q maps β onto itself.

<u>Proof</u>: Let $\beta \to \beta'$ then $q \in \beta'$. If β is a null Arc the assertion follows easily.

Assume β is a segment. Let $M(q)$ intersect the boundary of $Q(a_1,a_2)$ in two points r_1,r_2 and let $N(q)$ intersect this boundary in points s_1 and s_2. Assume without loss of generality $(a_1s_1a_3)$, $(a_3r_2a_2)$, $(a_2s_2a_4)$ and $(a_4r_1a_1)$ all hold. If $c_1 \in \alpha(a_1,r_1)$, then $c_2 \in \alpha(a_2,r_2)$. For $\beta' = \alpha(c_1',c_2')$ then $c_1' \in \alpha(a_2,r_2)$ and $c_2' \in \alpha(a_1,r_1)$.

Now $c_1q = c_1'q$ and $c_2q = c_2'q$. If $c_1' = c_2$ either $(c_1c_2'a_1)$ and $(c_2c_1'a_2)$ or else $(c_2'c_1a_1)$ and $(c_1'c_2a_2)$. Let the former hold. Then $qc_1 < qc_2' = qc_2 < qc_1'$ which is impossible.

This lemma implies that $R_{a_3a_4} \circ R_{a_1a_2} = R_{a_1a_2} \circ R_{a_3a_4} = R_q$.

In the set U_0 choose points a_1,a_2,b_1 and b_2 such that $a_1a_2 > 0$, $b_1b_2 > 0$, and $q = m_{a_1a_2} = m_{b_1b_2}$. If also

$$\alpha(b_1,b_2) \cap \alpha(a_1,a_2) = \alpha(b_1,b_2) \cap \alpha(a_3,a_4) = q$$

then set $b_3 = M(b_1) \cap N(b_2)$ and $b_4 = M(b_2) \cap N(b_1)$. Define

$$R_q = R_{a_3a_4} \circ R_{a_1a_2} \qquad \text{and} \qquad \bar{R}_q = R_{b_3b_4} \circ R_{b_1b_2}.$$

The two maps R_q and $\bar{R}_q$ agree on $\bar{Q}(a_1,a_2) \cap \bar{Q}(b_1,b_2)$. Any Arc β in $\bar{Q}(a_1,a_2) \cap \bar{Q}(b_1,b_2)$ that either intersects both $\alpha(a_1,a_2)$ and $\alpha(a_3,a_4)$ or else $\alpha(b_1,b_2)$ and $\alpha(b_3,b_4)$ is mapped isometrically onto an Arc β' under R_q and $\bar{R}_q$. For all distinct pairs r and s of points sufficiently close to q, the line $P(r,s)$ intersects either both $\alpha(a_1,a_2)$ and $\alpha(a_3,a_4)$ or both $\alpha(b_1,b_2)$ and $\alpha(b_3,b_4)$.

(VI.8) <u>Theorem</u>. Let R satisfy the bisector condition. Then for each $q \in R$ there is a neighborhood $W(q)$ and a topological isometry R_q of $\bar{W}(q)$ onto itself such that the only fixed point of R_q is q, and R_q^2 is the identity. Furthermore, any segment β in $W(q)$ that contains q and has its endpoints equidistant from q is mapped onto itself.

The set $W(q)$ is in general a proper subset of $Q(a_1,a_2) \cap Q(b_1,b_2)$. Without loss of generality, the neighborhood $W(q)$ can be chosen to be a null box. In the rest of this Chapter $W(q)$ is reserved to denote a null box satisfying (VI.8).

<u>Definition</u>: The general G-space R is called locally symmetric if, for each $p \in R$ there is a neighborhood $U(p)$ and a topological isometry T of $U(p)$ onto itself with p the only fixed point of T and with T^2 the identity.

Theorem (VI.8) says that a doubly timelike surface R satisfying the bisector condition is locally symmetric.

In the above neighborhood $W(q)$, consider points c_1 and c_2 with

(c_1qc_2) and $q = m_{c_1c_2}$. Define $c_3 = M(c_1) \cap N(c_2)$, $c_4 = M(c_2) \cap N(c_1)$ and $R_q' = R_{c_3c_4} \circ R_{c_1c_2} = R_{c_1c_2} \circ R_{c_3c_4}$. This is a topological isometry of $\bar{Q}(c_1,c_2)$. Furthermore, $R_{c_1c_2} = R_{c_3c_4} \circ R_q' = R_q' \circ R_{c_3c_4}$.

(VI.9) Theorem. $R_{c_1c_2}$ is a topological isometry of $\bar{Q}(c_1,c_2)$.

Proof: Let $x,y \in \bar{Q}(c_1,c_2)$ with $x \neq y$ and $x' = R_{c_1c_2}(x)$, $y' = R_{c_1c_2}(y)$. If $P(x,y) \cap \alpha(c_1,c_2) \neq \phi$ then (VI.6) guarantees $x'y' = xy$. Assume $P(x,y) \cap \alpha(c_1,c_2) = \phi$. Then $P(x,y) \cap \alpha(c_3,c_4) = r$. By (VI.6) $R_{c_3c_4}(x)\, R_{c_3c_4}(y) = xy$. Therefore, $x'y' = R_q' R_{c_3c_4}(x)\, R_q' R_{c_3c_4}(y) = xy$.

The map $R_{c_1c_2}$ will be called the reflection in $\alpha(c_1,c_2)$. By continuity of the metric, the points c_1 and c_2 can always be chosen on the boundary of $W(q)$. Consider a line P that is not null with $P \cap W(q) \neq \phi$. Let P intersect the boundary of $W(q)$ in points a_1 and a_2. Define R_P to be $R_{a_1a_2}$.

If $a_1a_2 > |\sigma| \geq 0$ define PT_σ as follows:
If $x \in \bar{Q}(a_1,a_2)$ let f denote the unique foot of x on $\alpha(a_1,a_2)$. Orient $P(a_1,a_2)$ such that a_2 follows a_1 if $\sigma \geq 0$ such that a_1 follows a_2 if $\sigma < 0$. Let $f(\sigma)$ denote the point (when it exists) that follows f on $\alpha(a_1,a_2)$ and has $ff(\sigma) = \sigma$. Choose $P_{f(\sigma)}$ to be the line perpendicular to P at $f(\sigma)$. Choose $x' \in P_{f(\sigma)}$ with $x'f(\sigma) = xf$ and such that x and x' lie on the same side of $\alpha(a_1,a_2)$. Then $PT_\sigma(x) = x'$.

If r and s are the points of $\alpha(a_1,a_2)$ with $a_1r = a_2s = |\sigma|$ then $PT_{|\sigma|}$ maps $\bar{Q}(a_1,r)$ topologically onto $\bar{Q}(s,a_2)$.

(VI.10) The map PT_σ is a topological isometry.

Proof: Let $q = \alpha(a_1,a_2) \cap \alpha(a_3,a_4)$. Consider $x,y \in \bar{Q}(a_1,r)$ and let x

and y have feet of f_1 and f_2 respectively on $\alpha(a_1,a_2)$. Let u_i be the midpoint of $\alpha(f_i, f_i(\sigma))$ for $i = 1,2$. Assume without loss of generality $qu_1 \geq qu_2 \geq 0$. Set $P_{u_2} \cap (\bar{W}(q) - W(q)) = \{c_1,c_2\}$ and $P_{f_2(\sigma)} \cap (\bar{W}(q) - W(q)) = \{d_1,d_2\}$. Then $x' = R_{d_1d_2} \circ R_{c_1c_2}(x)$ and $y' = R_{d_1d_2} \circ R_{c_1c_2}(y)$. Then, $x'y' = xy$.

(VI.11) Let $q_1,q_2 \in W(q)$, then there exist neighborhoods $U(q_1)$ and $U(q_2)$ that are topologically isometric.

Proof: If $q_1q_2 > 0$ let $P = P(q_1,q_2)$ and set $\sigma = q_1q_2$. Then PT_σ is a topological isometry of some neighborhood of q_1 onto some neighborhood of q_2.

If $q_1q_2 = 0$ choose $q_3 \in W(q)$ with $q_1q_3 > 0$ and $q_3q_2 > 0$. Then let $P_1 = P(q_1,q_3)$ and $P_2 = P(q_2,q_3)$. For $\sigma_i = q_iq_3$, where $i = 1,2$, the map $P_2T_{\sigma_2} \circ P_1T_{\sigma_1}$ is a topological isometry of some neighborhood of q_1 to some neighborhood of q_3.

(VI.12) Theorem. If R is a doubly timelike surface satisfying the bisector condition, any pair of points $p_1,p_2 \in R$ have neighborhoods that are topologically isometric.

Proof: Let S_1 be the set of points of R that have a neighborhood topologically isometric to some neighborhood of p_1. Then S_1 is open and nonempty by (VI.11).

Let S_2 be the compliment of S_1 in R. By (VI.11) the set S_2 is open. Since S_1 is then both open and closed, $S_1 = R$.

2. Angles. A direction from, or with, origin q, is an oriented Arc $\alpha^+(q,p)$ for p on the boundary of $W(q)$. The boundary of $W(q)$ is homeo-

morphic to an ordinary circle and each of its points lies on exactly one direction from q. An angle A with vertex q is a set of directions with origin q passing through the points of a subarc of the boundary of W(q). If u_1,u_2 are the endpoints of the arc, then the directions δ_i from q through u_i are the legs of the angle.

We distinguish three cases:

1.) If δ_1 and δ_2 are not opposite, then exactly one of the angles with legs δ_1 and δ_2 contains with some direction from q also the opposite direction. We call this angle the "concave angle" with legs δ_1,δ_2 and denote it by $(\delta_1,\delta_2)^{cav}$. The other angle is the "convex angle" with legs δ_1,δ_2 and denoted by $(\delta_1,\delta_2)^{vex}$. For convenience, we include the case where $\delta_1 = \delta_2$. In this case $(\delta_1,\delta_2)^{vex}$ consists of δ_1 alone, and $(\delta_1,\delta_2)^{cav}$ consists of all directions from q.

2.) If δ_1 and δ_2 are opposite, we call each of the angles with legs δ_1,δ_2 "straight".

3.) If $P(q,u_1) \perp P(q,u_2)$ we call $(\delta_1,\delta_2)^{vex}$ a right angle.

The null lines M(q) and N(q) through q divide W(q) into four open sectors S_1,S_2,S_3 and S_4. Assume without loss of generality that for $i = 1,2$ and $x \in S_i$ we have $q <_i x$. When δ_1 and δ_2 both lie in the same sector S_j, we say $(\delta_1,\delta_2)^{vex}$ is an angle of S_j.

Let $M^+(q)$ and $N^+(q)$ be orientations of M(q) and N(q). An orientation preserving rotation about q will mean an isometry of some neighborhood U(q) onto a neighborhood V(q) such that $U(q) \cap M(q)$ and $U(q) \cap N(q)$ are mapped respectively to $V(q) \cap M(q)$ and $V(q) \cap N(q)$ in such a way that the orientations of $M^+(q)$ and $N^+(q)$ are preserved. If α is a subarc of the direction δ at q and the orientation preserving rotation Φ maps α to α' a subarc of the direction δ', we say Φ rotates δ to δ'. If Φ rotates δ_i to δ_i', for $i = 1,2$, we say $(\delta_1,\delta_2)^{vex}$ has been rotated to $(\delta_1',\delta_2')^{vex}$.

(VI.13) Let R satisfy the bisector condition. If δ_1 and δ_2 are directions in S_j, there is an orientation preserving rotation Φ about q that rotates δ_1 into δ_2.

<u>Proof</u>: Choose $r_i \in \delta_i - q$, for $i = 1,2$, with $r_1 r_2 > 0$. Assume without loss of generality that $S_j = S_1$ and $r_1 <_2 r_2$. Let $P = P(q,m_{r_1 r_2})$ and define $\Phi = R_{P(q,r_2)} \circ R_P$. This is an orientation preserving rotation of q that rotates δ_1 to δ_2.

If A is the angle $(\delta_1,\delta_2)^{vex}$ and A_ν the angle $(\delta_1^\nu,\delta_2^\nu)^{vex}$ we say $A_\nu \to A$ if, and only if, the sets $(\delta_1^\nu,\delta_2^\nu)^{vex}$ tend in the sense of Hausdorff's closed limit to the sets $(\delta_1,\delta_2)^{vex}$.

An <u>angular</u> <u>measure</u> on the set of angles in S_j is a non-negative function $|A|$ defined for all angles in S_j such that

1.) $|A| = 0$ if, and only if, A consists of a single direction.

2.) If $A_1 \cap A_2$ consists of one direction, then $|A_1 \cup A_2|$ equals $|A_1| + |A_2|$.

If we require that $|A_1| = |A_2|$ when A_1 and A_2 are angles in S_j such that A_1 can be rotated into A_2, the angular measure is unique up to multiplication by a constant. Choose a measure $|A_1|_1$ for angles A_1 in S_1 that is invariant under rotations. Let $R_q(S_3) = S_1$. If A is an angle in S_3 then $R_q(A)$ is an angle in S_1. Define $|A| = |R_q(A)|_1$. If

δ_1, δ_2 are directions belonging to S_j, for $j = 2$ or 4, there are unique directions δ_1' and δ_2' in S_1 perpendicular to δ_1 and δ_2 respectively. Define $|(\delta_1,\delta_2)^{vex}| = |(\delta_1',\delta_2')^{vex}|$.

This gives an angular measure $|A|$ for all angles A at q that lie in some S_j, for $j = 1,2,3,4$. If q_1 is a second point of R, let T be a topological isometry of $U(q)$ onto $U(q_1)$ for $T(q) = q_1$. If δ is a direction at q and $\alpha \subset \delta$ is mapped under T to $\alpha' \subset \delta'$ for δ' a direction at q_1, we say T maps δ to δ'. Thus every angle A at q is mapped to an angle A' at q_1. If A belongs to S_j at q, then A' will belong to some sector S_i' at q_1. Define $|A'| = |A|$ for $T(A) = A'$. This definition is independent of the topological isometry T and gives a measure for all angles A' at q_1 belonging to sector S_j', for $j = 1,2,3,4$.

3. Lambert Quadrangles. Let w be a point of $W(q)$ and let P_1 and P_2 be two lines (neither null) through w with $P_1 \perp P_2$. If $p \in W(q)$ and if p is sufficiently near w, let x and y be the unique feet of p on P_1 and P_2 respectively. If $p \neq x$ set $P_2(p) = P(p,x)$ and if $p = x$ let $P_2(p)$ be the unique line perpendicular to P_1 at x. If $p \neq y$ let $P_1(p) = P(p,y)$ and if $p = y$, let $P_1(p)$ be the unique line perpendicular to P_2 at y.

There is a neighborhood H of w such that for all $p \in H$ the maps $R_{P_1(p)}$ and $R_{P_2(p)}$ are defined on (at least) all of H. We may take H to be a null box.

A convex quadrangle with three right angles is traditionally called a Lambert quadrangle. Therefore, we shall refer to the open set bounded by $\alpha(w,y)$, $\alpha(y,p)$, $\alpha(p,x)$ and $\alpha(x,w)$ as a Lambert quadrangle. In general abcd will denote a Lambert quadrangle with right angles at a,b,c and with d opposite b.

For each $p \in H$ define $px = \mu$, $py = \lambda$, $wx = \xi$ and $wy = \eta$.

The quantities ξ and η for example, determine μ and λ. We write $\mu = \mu(\xi,\eta)$ and $\lambda = \lambda(\xi,\eta)$. We only consider $p \in H_1$ where H_1 is a fixed component of $H - P_1 - P_2$. For $p_0 \in H_1$ let x_0 and y_0 be the feet of p_0 on P_1 and P_2 respectively. Then $\xi_0 = wx_0 > 0$, $\eta_0 = wy_0 > 0$, $\mu_0 = \mu(\xi_0,\eta_0)$ and $\lambda_0 = \lambda(\xi_0,\eta_0)$. One of the following must hold: $\xi_0 < \lambda_0$, $\xi_0 = \lambda_0$ or $\xi_0 > \lambda_0$.

(VI.14) If $\xi_0 = \lambda_0$, then $\eta_0 = \mu_0$.

Proof: Let $s = m_{wx_0}$ and $r = m_{wy_0}$. Then $R_{P_1(r)}$ maps $\alpha(w,x_0)$ onto $\alpha(y_0,p_0)$ since it preserves length and perpendicularity. This implies $\alpha(x_0,p_0)$ is mapped onto itself and that $P_1(p_0) \perp P_2(p_0)$.

$R_{P_2(s)}$ maps (at least part of) $P_2(p_0)$ to P_2. The foot of $s' = P_2(s) \cap \alpha(y_0,p_0)$ on $P_2(p_0)$ must be mapped to the foot of s' on P_2. But this means $R_{P_2(s)}(p_0) = y$. Consequently, $p_0x_0 = y_0w_0$ and $\eta_0 = \mu_0$.

Assuming $\xi_0 = \lambda_0$ consider (ξ,η) for $0 < \xi < \xi_0$ and $0 < \eta < \eta_0$ as cartesian coordinates on the Lambert quadrangle $x_0wy_0p_0$. Let $p_1 = (2^{-1}\xi_0, 2^{-1}\eta_0)$. The map $R_{P_1}(p_1)$ interchanges $\alpha(w,x_0)$ and $\alpha(y_0,p_0)$. Consequently, $R_{P_1}(p_1)$ interchanges m_{wx_0} and $m_{y_0p_0}$. Yet $R_{P_2(p_1)}$ maps $\alpha(y_0,p_0)$ on itself with y_0 and p_0 being interchanged. Therefore, $m_{y_0p_0} \in P_2(p_1)$. In the same way one can show $m_{wx_0} \in P_2(p_1)$. This implies $p_1 \in P(m_{wx_0}, m_{y_0p_0})$ and consequently, $P_1(p_1) \perp P_2(p_1)$. The four Lambert quadrangles formed by the components of $x_0wy_0p_0 - P_1(p_1) - P_2(p_1)$ all have four right angles. By continuing with this type of argument one can deduce the following result.

(VI.15) If $\xi_0 = \lambda_0$ and $p \in x_0wx_0p_0$, then $P_1(p) \perp P_2(p)$.

A general G-space is called <u>locally Minkowskian</u> if each point has a neighborhood topologically isometric to an open set in a Minkowski space.

(VI.16) <u>Theorem</u>. Let R satisfy the bisector condition and let one Lambert quadrangle $x_0wy_0p_0$ in H have two opposite sides equal. Then R is locally Minkowskian.

<u>Proof</u>: Let p be an interior point of $x_0wy_0p_0$ and let α be an Arc through p with endpoints u_1, u_2 on the boundary of $x_0wy_0p_0$.

First, we wish to show α has a linear representation $A\xi + B\eta + C = 0$ in the (ξ,η) coordinates. Assume $u_1 \notin P_1(p) \cup P_2(p)$ since otherwise the problem is trivial. It is only necessary to show that if $u_1 = (\xi_1,\eta_1)$ and $p = (\xi_2,\eta_2)$ the point $v = 2^{-1}(\xi_1 + \xi_2, \eta_1 + \eta_2)$ lies on $\alpha(p,u_1)$. Notice that if $v_1 = P_1(p) \cap P_2(u_1)$ and $v_2 = P_2(p) \cap P_1(u_1)$ the Lambert quadrangle $pv_1u_1v_2$ has four right angles. Let $\alpha(p,u_1) \cap P_i(v) = c_i$ for $i = 1,2$. Then $R_{P_i(v)}$ takes $\alpha(p,u_1)$ to $\alpha(v_1,v_2)$ and maps c_i to c_i. Therefore, c_1 and c_2 lie on $\alpha(v_1,v_2)$ which implies $c_1 = c_2 = v$.

Notice that R_v, the reflection in v, interchanges p and u_1 so that $pv = u_1v$. From this it follows that the metric midpoint of a segment α in $x_0wy_0p_0$ is the cartesian midpoint in terms of the (ξ,η) coordinates.

Still using the same notation, let $p \in x_0wy_0p_0$ and consider the tangent space T_p. Clearly, the tangent space exists and the metric in T_p agrees with the original metric in $x_0wy_0p_0$. Since T_p is Minkowskian and satisfies the bisector condition locally, it satisfies the bisector condition in the large. That is, if $a_1, a_2 \in T_p$ with $a_1a_2 > 0$, the bisector $B(a_1,a_2)$ contains a line S in T_p such that $S \cap \alpha(a_1,a_2)$ is a single point. By the same type of arguments as in Section 2 of this Chapter, we can show T_p may be reflected in any line not a null line.

(VI.17) <u>Theorem</u>. The unit sphere K in T_p consists of two quadrics. Each quadric is a dilation of the conjugate of the other.

If the bisector condition is slightly strengthened, one can establish the following result.

(VI.18) <u>Theorem</u>. Let R be a doubly timelike surface. For each $q \in R$ let there exist a neighborhood U_0 in which all distinct pairs of points a_1, a_2 have a bisector $B(a_1, a_2)$ with a locally flat subset S. Then each point of R has a neighborhood topologically isometric to an open set of the two dimensional Lorentz space.

Consider now the situation when $\xi_0 \neq \lambda_0$.

(VI.19) If $\xi_0 < \lambda_0$, then $\eta_0 > \mu_0$. If $\xi_0 > \lambda_0$, then $\eta_0 < \mu_0$.

<u>Proof</u>: The proof of each case is similar, so we will only show $\xi_0 < \lambda_0$ implies $\eta_0 > \mu_0$.

Because of (VI.16), $\xi_0 < \lambda_0$ implies $\xi < \lambda(\xi, \eta)$ for all $0 < \xi \leq \xi_0$ and $0 < \eta \leq \eta_0$. We may assume then, without loss of generality, that $w \in M(p_0)$ and hence $wp_0 = 0$. Let r be the midpoint of $\alpha(w, y_0)$. $R_{P_1(r)}$ maps (at least part of) $P_1(p_0)$ to P_1. Since $y_0 p_0 > wx_0$ we have $(wx_0 R_{P_1(r)}(p_0))$. But this implies that if $u = N(x_0) \cap P_1(p_0)$ then $(uy_0 p_0)$. Let $s = m_{wx_0}$. Then $R_{P_2(s)}(p_0) = u$. Since y_0 is the foot of w on $P_1(y_0)$ we have $wy_0 > wu = x_0 p_0$. Therefore, $\eta_0 > \mu_0$.

By renaming P_1 and P_2 (if necessary), we may assume then that $\xi_0 < \lambda_0$ and $\eta_0 > \mu_0$ in the rest of this chapter. Without loss of generality, assume $w <_1 x_0$ and $w <_2 y_0$. Then any Lambert quadrangle $abcd$ in H with $b <_1 a$ has $ab < cd$ and $bc > ad$.

4. <u>Equidistant Curves</u>. Let $\alpha = \alpha(a_1,a_2)$ be a segment with endpoints a_1,a_2 on the boundary of H. Let Q_1 and Q_2 be the components of $\bar{Q}(a_1,a_2) - \alpha(a_1,a_2)$. Each $p \in \bar{Q}_i$ has exactly one foot f on $\alpha(a_1,a_2)$. Define $E_i(\Theta) = \{p \in \bar{Q}_i | pf = \Theta\}$. For sufficiently small $\Theta \geq 0$, the set $E_i(\Theta)$ will consist of a simple Jordan arc connecting two points of the boundary of Q_i. We call $E_i(\Theta)$ the equidistant curve of distance Θ from $\alpha(a_1,a_2)$ with respect to Q_i.

If Θ is positive yet so small that $E_i(\Theta)$ consists of more than one point, $Q_i - E_i(\Theta)$ has two components. If for all distinct pairs $x,y \in E_i(\Theta)$, we have $p \in \alpha(x,y) - x - y$ implies $pf > \Theta$, we call $E_i(\Theta)$ strictly convex. If $p \in \alpha(x,y) - x - y$ always implies $pf < \Theta$ we call $E_i(\Theta)$ strictly concave.

Let $P = P(a_1,a_2)$ and $a_1a_2 > |\sigma| \geq 0$. Then PT_σ maps all of $E_i(\Theta)$ in its domain onto all of $E_i(\Theta)$ in its range. Any point x in the relative interior of $E_i(\Theta)$ can be mapped by some PT_σ to any other point y in the relative interior of $E_i(\Theta)$. Thus, if $E_i(\Theta)$ is strictly convex (concave) near one point it is strictly convex (concave).

If $a_1 <_i a_2$ for $i = 1,2$ there is an orientation preserving rotation about $q = \alpha(a_1,a_2) \cap \alpha(a_3,a_4)$ taking (at least part of) $\alpha(a_1,a_2)$ to (at least part of) $P_i(q)$.

(VI.20) If $a_1 <_1 a_2$ then $E_1(\Theta)$ and $E_2(\Theta)$ are strictly convex. If $a_1 <_2 a_2$ then $E_1(\Theta)$ and $E_2(\Theta)$ are strictly concave.

<u>Proof</u>: We only consider $a_1 <_1 a_2$. Assume $\alpha(a_1,a_2)$ intersects $\alpha(a_3,a_4)$ in w and $\alpha(a_1,a_2) \subset P_1$.

Let p_1,p_2 be distinct points of $E_i(\Theta)$ and $p_3 \in \alpha(p_1,p_2) - p_1 - p_2$. Let p_j have foot f_j on P_1 for $j = 1,2,3$. Set $P_1(p_3) \cap P_2(p_k) = s_k$

for $k = 1,2$. If $p_3f_3 \leq \Theta$, then (VI.19) implies $(p_1s_1f_1)$ and $(p_2s_2f_2)$. But then $p_3 \in \alpha(p_1,p_2) \cap \alpha(s_1,s_2)$ which is impossible.

(VI.21) $E_i(\Theta)$ is differentiable.

Since $E_i(\Theta)$ is a local T-curve, it is rectifiable with length $L(E_i(\Theta))$. Any subarc of $E_i(\Theta)$ can be mapped to any other subarc having the same length by some PT_σ. Therefore, for fixed Θ there is a number $\phi_j(\Theta)$ such that for $p_1,p_2 \in E_i(\Theta)$ with feet f_1 and f_2 on $\alpha(a_1,a_2)$ we have $L(A(p_1,p_2)) = \phi_j(\Theta)f_1f_2$ where $A(p_1,p_2)$ is the arc of $E_i(\Theta)$ from p_1 to p_2.

There are actually two functions $\phi_1(\Theta)$ and $\phi_2(\Theta)$ corresponding to $a_1 <_1 a_2$ and $a_1 <_2 a_2$ respectively. The process of determining $\phi_j(\Theta)$ is the same in both cases, so we will only discuss in detail the case $\phi_1(\Theta)$.

Using the same notation as previously discussed, let P_1 and P_2 be lines perpendicular at w. For $0 < \Theta < \mu_0$, define $p_\Theta(t) = (\Theta,t)$ where $0 \leq t \leq \xi_0$. Then $p_\Theta(t)$ is an equidistant curve from P_1 of length $\phi_1(\Theta)\xi_0$. Assume $p_\Theta(0) \in E_1(\Theta)$ and denote the image of $p_\Theta(t)$ by $E'(\Theta)$. Then $E'(\Theta) \subset E_1(\Theta)$.

(VI.22) $\phi_1(\Theta)$ is a continuous function of Θ.

<u>Proof</u>: Let $\Theta_\nu \to \Theta^+$, then each $p \in E'(\Theta_\nu)$ has a foot $f \in P_1$ and $pf > \Theta$. Therefore, (III.15) implies $L(E'(\Theta_\nu)) \to L(E'(\Theta))$.

Let $\Theta \to \Theta^-$. For each fixed ν there are numbers γ such that all $p \in P_2T_\gamma(E'(\Theta_\nu))$ have feet f on P_1 with $pf \geq \Theta$. Let γ_ν be the infimum of these γ. Then $\gamma_\nu \to 0^+$. Furthermore, $P_2T_{\gamma_\nu}(E'(\Theta_\nu)) \to E'(\Theta)$ implies $L(E'(\Theta_\nu)) \to L(E'(\Theta))$ because of (III.15).

The above proof can be modified to show the next result.

(VI.23) If C_ν is a sequence of convex (concave) local T-curves converging to the equidistant convex (concave) curve $E_i(\Theta)$ then $L(C_\nu) \to L(E_i(\Theta))$.

For $0 < \Theta < \mu_0$ choose $h > 0$ such that $0 < \Theta-h < \Theta < \Theta+h < \mu_0$. Using (ξ,η) coordinates define $r_0^n = (0,0)$, $r_1^n = (2^{-1}n^{-1}\xi_0,0)$, $r_2^n = (n^{-1}\xi_0,0)$, $u_i^n = P_2(r_i^n) \cap E'(\Theta-h)$, $v_i^n = P_2(r_i^n) \cap E'(\Theta)$ and $w_i^n = P_2(r_i^n) \cap E'(\Theta+h)$ for $i = 0,1,2$. If $p,q \in E'(\Theta)$ let $A_\Theta(p,q)$ denote the arc of $E'(\Theta)$ from p to q. For fixed n let $s_0 = m[r_0^n,r_1^n] = m_{r_0^n r_1^n}$, $s_1 = P_2(s_0) \cap \alpha(u_0^n,u_1^n)$, $s_2 = P_2(s_0) \cap \alpha(v_0^n, v_1^n)$ and $s_3 = P_2(s_0) \cap \alpha(w_0^n, w_1^n)$.

(VI.24) $P_2(s_0)$ is $\perp$ to all three of $P(u_0^n, u_1^n)$, $P(v_0^n, v_1^n)$ and $P(w_0^n, w_1^n)$.

If L_i is the line $\perp$ to $P(v_{i-1}^n, v_i^n)$ at v_i^n let $s_4 = L_1 \cap \alpha(w_0^n, w_1^n)$ and $s_5 = L_2 \cap \alpha(w_1^n, w_2^n)$. Then $(s_3 s_4 w_1^n)$ and $(s_6 s_5 w_2^n)$ for $s_6 = \alpha(w_1^n, w_2^n) \cap P_2(m[v_1^n,v_2^n])$.

The convex angle formed by $\alpha(v_1^n, s_4)$ and L_1 is equal to the convex angle formed by $\alpha(v_2^n, s_5)$ and L_2.

Consider the map $R_n = R_{P_1(v_1^n)}$ for fixed n. This map interchanges u_1^n and w_1^n. Let α' denote the image of the Arc α under R_n. Let x' denote the image of the point x under R_n. Then $\alpha'(v_1^n, v_2^n) \subset P(v_0^n, v_1^n)$.

(VI.25) $P(w_0^n, w_1^n)$ intersects $\alpha'(v_2^n, u_2^n)$ in a relative interior point.

<u>Proof:</u> $v_1^n s_2 s_3 s_4$ is a Lambert quadrangle and so is $(v_2^n)' s_2 s_3 s_7$ for $s_7 = P((v_2^n)', s_5') \cap P(w_0^n, w_1^n)$. Let $h_1 = v_1^n s_4$ and $h_2 = (v_2^n)' s_7$. Then $h_1 > h_2$. But this implies $h = v_1^n w_1^n$ is larger than the distance from $(v_2^n)'$

to $P(w_0^n, w_1^n) \cap P((v_2^n)', (u_2^n)')$ since the convex angles $\alpha(v_1^n, s_4)$ with L_1 and $\alpha((v_2^n)', (u_2^n)')$ with $\alpha((v_2^n)', s_5')$ are equal.

In the same way one can show $P(w_1^n, (u_2^n)') \cap P_2$ is a relative interior point of $\alpha(v_0^n, w_0^n)$.

Define F_0^n to be the compact set bounded by $\alpha(u_0^n, w_0^n)$, $\alpha(w_0^n, w_1^n)$, $\alpha'(u_1^n, u_2^n)$, $\alpha'(u_2^n, w_2^n)$, $\alpha'(w_2^n, w_1^n)$ and $\alpha(u_1^n, u_0^n)$. Let $P = P(v_0^n, v_1^n)$ and set $F_k^n = PT^k_{n^{-1}\mu_0}(F_0^n)$ for $k = 1,2,\cdots,n-1$. For large n all the sets F_k^n will exist. It can be shown that $F_{k-1}^n \cap F_k^n$ is the segment $PT^k_{n^{-1}\mu_0}(\alpha(u_0^n, w_0^n))$.

The Arcs $\alpha(u_1^n, w_1^n)$ and $\alpha'(u_2^n, w_2^n)$ are not $\perp$ to P, but as $n \to \infty$ they converge to $\alpha(u_0^n, w_0^n)$ which is $\perp$ to P.

The set

$$F = \lim_{n\to\infty} \bigcup_{k=0}^{n-1} F_k^n$$

exists and is not empty. Let P intersect the boundary of F in v_0^1 and z_0. Then $v_0^n z_0 = v_0^1 z_0 = \phi_1(\Theta)\xi_0$. The set F is bounded by two Arcs α_1 and α_2' perpendicular to P at v_0^1 and z_0 and by two equidistant curves from P at distance h. The total boundary of F has length $2\phi_1(\Theta)\xi_0\phi_1(h) + 4h$. The length of the boundary of the set $\bigcup_{k=0}^{n-1} F_k^n$ is $4h + n(2u_0^n u_1^n + 2w_0^n w_1^n)$. Furthermore,

$$\lim_{n\to\infty} \{4h + n(2u_0^n u_1^n + 2w_0^n w_1^n)\} = 4h + \phi_1(\Theta-h)\mu_0 + \phi_1(\Theta+h)\xi_0.$$

(VI.26) $2\phi_1(\Theta)\phi_1(h) = \phi_1(\Theta-h) + \phi_1(\Theta+h)$.

<u>Proof</u>: We must show that the length of the boundary of $\bigcup_{k=0}^{n-1} F_k^n$ converges to the length of the boundary of the set F.

Because of (VI.23) it is only necessary to show that

$$\bigcup_{k=0}^{n-1} PT^{k}_{n^{-1}\mu_0}(\alpha(w_0^n, w_1^n) \cup \alpha'(u_1^n, u_2^n))$$

is a convex curve. This follows from (VI.25) and the remark immediately after. Then $\phi_1(\Theta) = \cosh k_1\Theta$, $\phi_2(\Theta) = \cos k_2\Theta$, see [3, p. 326].

5. The Pythagorean Theorem. We now add the assumption that R is differentiable at w with respect to some segment α. Let w, H, P_1, etc. be as before and denote the components of $H - M(w) - N(w)$ by S_1, S_2, S_3 and S_4. Assume without loss of generality that $R_w(S_1) = S_3$ and $P_1 \cap H \subset S_1 \cup S_3 \cup \{w\}$.

(VI.27) If $\alpha \subset S_i \cup S_{i+2} \cup \{w\}$ for $i = 1$ or 2, then R is differentiable with respect to any segment β having w in its relative interior and such that $\beta \subset S_i \cup S_{i+2} \cup \{w\}$.

Proof: There is an orientation preserving rotation about w taking (at least part of) α to (at least part of) β.

Consider $p \in x_0wy_0p_0 - N(w) - Q(w)$. Let A and B denote respectively the convex angles at w determined by $\alpha(w,p)$ with P_1 and $\alpha(w,p)$ with P_2. If $p \in S_1$, then A is an angle in S_1. If $p \in S_2$, then A contains a null direction and B is an angle in S_2.

As before, we let x and y denote the feet of p on P_1 and P_2 respectively. Also,

$$\rho = wp, \quad \xi = wx \quad \text{and} \quad \eta = wy.$$

Given A and ξ (or η) the number ρ is determined so we write $\rho = \rho(A,\xi)$ and $\rho = \rho(A,\eta)$.

(VI.28) If A is a fixed angle in S_1, then $\rho(A,\xi)$ is an increasing strictly convex function of ξ. If A contains a null direction, then $\rho(A,\eta)$ for fixed A is an increasing strictly concave function of η.

Proof: The proofs are similar so we only consider the case of A, an angle in S_1.

Let $0 < \xi-h < \xi < \xi+h < \xi_0$. In (ξ,η) coordinates let $x_1 = (\xi-h,0)$ and $x_2 = (\xi+h,0)$. Set $q_i = P_2(x_i) \cap P(w,p)$ and $a_i = P_1(p) \cap P_2(x_i)$ for $i = 1,2$. Let L_1 and L_2 be the lines perpendicular to $P_1(p)$ at a_1 and a_2 respectively. Then $(L_i \cap P_1)xpa_i$ are Lambert quadrangles and hence $(L_i \cap P_i)x > h$. Therefore, if $r_i = L_i \cap P(w,p)$ we have (r_1q_1p) and (pr_2q_2). Yet $pr_1 = pr_2$, $pq_1 < pq_2$. Hence, $2\rho(A,\xi)$ is less than $\rho(A,\xi-h) + \rho(A,\xi+h)$.

Assume without loss of generality that H is contained in the neighborhood $V(w)$ with

$$\frac{1}{2} \le \frac{m(a,b)}{ab} \le \frac{3}{2} \quad \text{for all } a,b \in V(w) \text{ with } ab > 0.$$

When $wp > 0$ define

$$C(A) = \lim_{\xi\to 0+} \frac{\xi}{\rho(A,\xi)}, \qquad C(B) = \lim_{\eta\to 0+} \frac{\eta}{\rho(A,\eta)},$$

$$S(A) = \lim_{\xi\to 0+} \frac{\mu(A,\xi)}{\rho(A,\xi)}, \qquad S(B) = \lim_{\xi\to 0+} \frac{\lambda(A,\xi)}{\rho(A,\xi)},$$

$$T(A) = C^{-1}(A)S(A) \quad \text{and} \quad T(B) = C^{-1}(B)S(B).$$

These limits exist because of the differentiability assumption.

If $p \in S_1$ let $b_1 = P_1 \cap M(p)$ and $b_2 = P_1 \cap N(p)$. If $p \in S_2$ let $c_1 = P_2 \cap M(p)$ and $c_2 = P_2 \cap N(p)$. Using the notation of Section 6, Chapter III, define $p(\Theta) = p(\Theta,\alpha,w)$, $b_i(\Theta) = b_i(\Theta,\alpha,w)$ and $c_i(\Theta) = c_i(\Theta,\alpha,w)$ for $i = 1,2$ and $0 < \Theta \leq 1$.

Because $m(a(\Theta),b(\Theta)) = \Theta m(a,b)$ we have $b_1(\Theta) = P_1 \cap M(p(\Theta))$, $b_2 = P_1 \cap N(p(\Theta))$, $c_1(\Theta) = P_2 \cap M(p(\Theta))$ and $c_2(\Theta) = P_2 \cap N(p(\Theta))$. Let $x(p(\Theta))$ and $y(p(\Theta))$ denote the feet of p on respectively P_1 and P_2.

(VI.29) If $p \in S_1$, then $C(A) = \dfrac{m(w,b_1) + m(w,b_2)}{2m(w,p)}$. If $p \in S_2$, then $C(B) = \dfrac{m(w,c_1) + m(w,c_2)}{2m(w,p)}$.

Proof: We will only show the result for $p \in S_1$ since the proofs are similar.

Let $\xi = wx(p(\Theta)) = \frac{1}{2}(wb_1(\Theta) + wb_2(\Theta))$. Then $C(A) = \lim_{\xi\to 0+} \dfrac{\xi}{\rho(A,\xi)}$ $= \lim_{\Theta\to 0+} \dfrac{(1/2)(wb_1(\Theta) + wb_2(\Theta))}{wp(\Theta)}$. Therefore, $C(A) = \dfrac{m(w,b_1) + m(w,b_2)}{2m(w,p)}$.

This lemma implies that $C(A)$ is finite, non-zero and a continuous function of A for $p \in S_1$. Actually $C(A)$ depends on the measure of the angle A for A belonging to S_1. Hence, we write $C(|A|)$.

(VI.30) Let R be differentiable and satisfy the bisector condition. If $\eta_0 > \eta_1 > 0$, then $\phi_1(\eta_1) = \lim_{\xi\to 0+} \dfrac{\lambda(\xi,\eta_1)}{\xi}$. If $\xi_0 > \xi_1 > 0$, then $\phi_2(\xi_1) = \lim_{\eta\to 0+} \dfrac{\mu(\xi_1,\eta)}{\eta}$.

Proof: We will show only the first assertion. Let $E_1(\eta_1)$ be an equidistant curve to P_1 at distance η_1. Let $y = E_1(\eta_1) \cap P_2$ and $p \in E_1(\eta_1) - P_2$. Let x be the foot of p on P_1 and set $q = P_1(y) \cap P_2(x)$. Then $P(y,q) = P_2(y)$ and $P(y,p) \to P_2(y)$ as $p \to y$. Since $E_1(\eta_1)$ is differentiable at y with tangent $P_2(y)$, we have $yp/yq \to 1$ as $p \to y$. Thus,

$$\phi_1(\eta_1) = \lim_{p\to y} \frac{py}{wx} = \lim_{p\to y} \frac{qy}{wx} = \lim_{\xi\to 0+} \frac{\lambda(\xi,\eta_1)}{\xi}.$$

The following can be shown without much difficulty.

(VI.31) If $\{p_\nu\}$ is a sequence of points in the interior of $x_0wy_0p_0$ with $p_\nu \to w$ and $p_\nu = (\xi_\nu, \eta_\nu)$, then

$$\frac{\lambda(\xi_\nu, \eta_\nu)}{\xi_\nu} \to 1 \quad \text{and} \quad \frac{\mu(\xi_\nu, \eta_\nu)}{\eta_\nu} \to 1.$$

(VI.32) If $p \in S_1 \cup S_2$, then $C(B) = S(A)$ and $S(B) = C(A)$.

The Pythagorean theorem for right triangles in H can now be deduced. Let $a,b,c \in H$ with $ab > 0$, $ac > 0$, $cb > 0$ and $P(b,c) \perp P(a,c)$.

(VI.33) <u>Pythagorean Theorem</u>. Assume $a <_1 c$ and $c <_2 b$. Then

$$\cos k_2ab = \cos k_2ac \cosh k_1bc \quad \text{if } a <_1 b$$

$$\cosh k_1ab = \cos k_2ac \cosh k_1bc \quad \text{if } a <_2 b$$

<u>Proof</u>: The proof is basically the same as found in [3, pp. 327-328].

Without loss of generality assume that the above triangle Δabc has its vertex a at w, b lies in $x_0wy_0p_0$ and $\alpha(a,c)$ lies on P_1. Let A be the angle determined at a by $\alpha(a,b)$ and $\alpha(a,c)$. Let B be the angle at a determined by $\alpha(a,y)$ and $\alpha(a,b)$ where y is the foot of b on P. Using the same type of arguments as in [3, p. 328] the next result can be established.

(VI.34) If $b \in S_1$ then A belongs to S_1 and

$$C(A) = \frac{\tan k_2ac}{\tan k_2ab}.$$

If $b \in S_2$ then B belongs to S_2 and

$$C(B) = \frac{\tanh k_1 ay}{\tanh k_1 ab} .$$

We wish to identify $C(|A|)$ with $\cosh k|A|$. Let $0 < h < |A|$ and A belong to S_1 with vertex w. Let A have legs R and R_1 with R lying on P_1. Construct the two angles D_1 and D_2 at w belonging to S_1 each with one leg R_1 and measure h. The perpendicular L to R_1 at a point y of R_1 close to w intersects R in a point q, and the other legs of the two angles D_1 and D_2 at points x and x' with $(qx'y)$. Denote the feet of x,y,x' on R by u,v,u'. Then let L' be the perpendicular to $P(u',x')$ at x'. Let $\bar{v}$ and $\bar{u}$ be the intersections of L' with $P(y,v)$ and $P(x,u)$. As $y \to w$ the lines L and L' converge respectively to the perpendicular at w to R_1 and to R. Let $D(x')$ be the angle at x' formed by $\alpha(x',y)$ and $\alpha(x',\bar{v})$. Then $D(x')$ converges to an angle D at w as $y \to w$. Furthermore, D has legs that are not null so $C(D)$ exists and is finite and non-zero.

$$C(D) = \lim_{y\to w} \frac{x'\bar{v}}{x'y} = \lim_{y\to w} \frac{x'\bar{u}}{x'x} .$$

Therefore, $\frac{x'\bar{u}}{x'\bar{v}}$ converges to two as $y \to w$. By (VI.31) we then have $\frac{u'v}{uv} \to 1$ as $y \to w$. Furthermore, $\frac{wv}{wx} = \frac{wv}{wy}\frac{wy}{wx} \to C(|A|)C(h)$ and

$$\frac{2wv}{wx} = \frac{wv}{wx} + \frac{wv}{wx'} = \frac{wu'}{wx'} + \frac{wu}{wx} + \frac{u'v-uv}{uv}\frac{uv}{wx'} .$$

But,

$$\frac{uv}{u'v} \to 1 \quad \text{and} \quad \frac{uv}{wx'} < \frac{wu}{wx'} .$$

Therefore, $C(|A| + h) + C(|A| - h) = 2C(|A|)C(h)$. Since $C(|A|) \geq 1$ and is not constant, we have $C(|A|) = \cosh k|A|$ for A belonging to S_1.

Using the same points w and y, let r be the foot of y on R. Then $\cos k_2 wy = (\cos k_2 wr)(\cosh k_1 yr)$. Therefore

$$S(|A|) = \lim_{y \to w} \frac{yr}{wy} = \lim_{y \to w} \frac{\cosh^{-1}\left[\dfrac{\cos k_2 wy}{\cos k_2 wr}\right]}{k_1 wy}$$

However,

$$C(|A|) = \frac{\tan k_2 wr}{\tan k_2 wy} \quad \text{so that}$$

$$S(|A|) = \lim_{y \to w} \frac{\cosh^{-1}[(\cos k_2 wy)(1 + C^2(A)\tan k_2 wy)^{1/2}]}{k_1 wy}$$

$$S(|A|) = \frac{k_2}{k_1} \cdot \lim_{x \to 0+} \frac{\cosh^{-1}[(\cos^2 x + C^2(A)\sin^2 x)^{1/2}]}{x}$$

$$S(|A|) = \frac{k_2}{k_1} \cdot (C^2(A) - 1)^{1/2} = \frac{k_2}{k_1} \sinh k|A|$$

(VI.35) If $p \in S_1$, then $S(B) = C(A) = \cosh k|A|$, $C(B) = S(A) = k_1^{-1}k_2 \sinh k|A|$ and $T(A) = k_1^{-1}k_2 \tanh k|A|$.

Consider now a triangle Δabc in H with $a <_1 b$, $a <_1 c$ and $b <_1 c$. Assume we are given ab, ac and the measure $|A|$ of the angle A at a determined by $\alpha(a,b)$ and $\alpha(a,c)$.

The distance bc may then be calculated using formulas we have derived. Let d be the foot of b on $P(a,c)$, then (adc).

$$ad = k_2^{-1}\tan^{-1}(C(|A|)\tan k_2 ab) \quad \text{by (VI.34)}$$

$$bd = k_1^{-1}\cosh^{-1}(\cos k_2 ab \sec k_2 ad) \quad \text{by (VI.33)}$$

$cd = ac - ad$

and $\quad bc = k_2^{-1} \cos^{-1}(\cos k_2 cd \cosh k_1 bd)$ by (VI.33)

Let Δabc lie in H with $a <_1 b$, $a <_1 c$ and $c <_2 b$. In the same fashion as above we may calculate bc given ab, ac and the angle A determined by $\alpha(a,b)$ and $\alpha(a,c)$.

We now wish to show that given a space R satisfying the bisector condition and differentiable at one point, the formulas (VI.33), (VI.34) and (VI.35) uniquely determine the distance bc of any two points b and c sufficiently close.

Let $a = w$ and choose $p \in H$ with $a <_1 p$. Let $U(p)$ be a neighborhood of p such that all $q \in U(p)$ satisfy $q \in H$ and $a <_1 q$. It is sufficient to show that given $b,c \in U(p)$ together with their (ξ,μ) coordinates the distance bc is determined.

The distances ab and ac can be determined by (VI.33). The angles that $\alpha(a,b)$ and $\alpha(a,c)$ make with P_1 can be determined by (VI.34) and (VI.35). Consequently, the angle A between $\alpha(a,b)$ and $\alpha(a,c)$ can be determined. Therefore, by the remarks after (VI.35) the distance bc can be calculated.

The space R may now be (locally) identified with $EH(k_1,k_2)$ since this space is differentiable and satisfies the bisector condition as well as the required formulas.

(VI.36) <u>Theorem</u>. Let R satisfy the bisector condition and be differentiable at some point. Then if R is not locally Minkowskian, there exist k_1 and k_2 such that any point of R has a neighborhood topologically isometric to an open set of $EH(k_1,k_2)$.

We have shown in this chapter that a doubly timelike surface R satisfying the bisector condition is locally symmetric and may be reflected (locally) in segments. For any two points p,q ε R there are neighborhoods U(p) and U(q) that are topologically isometric.

Theorem (VI.16) gives a condition that guarantees R is locally Minkowskian. Theorem (VI.17) states that when R satisfies the bisector condition and is locally Minkowskian it is nearly locally Lorentzian. R is locally Lorentzian exactly when (locally) it may be reflected in null lines, see (VI.18).

Theorem (VI.36) says R is locally exterior hyperbolic if it satisfies the bisector condition, is differentiable at some point, and is not locally Minkowskian.

REFERENCES

1. A. D. Alexandorov, Über Eine Verallgemeinerung der Riemannschen Geometrie, Ber. Reimann-Tagung Forschungsinst. fur Math., Heft 1(1957), 33-84.

2. J. K. Beem, Synthetic Theory of Indefinite Metric Spaces, Dissertation, University of Southern California, 1968.

3. H. Busemann, The Geometry of Geodesics, Academic Press, New York, 1955.

4. H. Busemann, Similarities and differentiability, Tôhoku Math. J. 9 (1957), 56-67.

5. H. Busemann, Timelike Spaces, Dissertationes Mathematicae 53, Warszawa (1967).

6. H. Busemann, Metric Methods in Finsler Spaces, Princeton Univ. Press, 1942.

7. H. Busemann and J. K. Beem, Axioms for Indefinite Metrics, Rend. Circolo Mat. Palermo, (1966), 223-246.

8. H. Busemann and P. J. Kelly, Projective Geometry and Projective Metrics, Academic Press, New York, 1953.

9. S. Cohn-Vossen, Existenz kürzester Wege, Dokl. Akad. Nauk SSSR 3 (1935), 339-342.

10. H. Freudenthal, Das Helmholtz-Liesche Raumproblem bei indefiniter Metrik, Math. Ann. 156 (1964), 263-312.

11. J. G. Hocking and G. Young, Topology, Addison-Wesley, Reading, Mass., 1961.

12. H. Hopf and W. Rinow, Über den Begriff der vollständigen differentialgeometrischen Fläche, Comment. Math. Helv. 3 (1932), 209-225.

13. W. Kaplan, Regular Curve-Families filling the plane.
I. Duke Math. Journal 7 (1940), 154-185.
II. Duke Math. Journal 8 (1941), 11-46.

14. D. Montgomery, Locally Homogeneous spaces, Ann. of Math. 52 (1950), 261-327.

15. M. Morse, A special parametrization of curves, Bull. Amer. Math. Soc. 42 (1936), 915-922.

16. W. Noll, Euclidean and Minkowskian geometry, Amer. Math. Monthly 71 (1964), 129-144.

17. L. S. Pontryagin, Theory of Groups, Gordon and Breach, New York, 1966.

18. W. Rinow, Die innere Geometrie der metrischen Räume, Berlin-Gottingen-Heidelberg, 1961.

19. B. A. Rozenfel'd, Non-euclidean geometries (Russian) Moscow, 1955.

20. H. Seifert and W. Threlfall, Lehrbuch der Topologie, Chelsea Press, New York, 1934.

21. H. E. Vaughan, On locally compact metrizable spaces, Bull. Amer. Math. Soc. 43 (1937), 532-535.

22. P. Woo, Doubly-Timelike General G-Spaces, Dissertation, University of Southern California, 1968.

23. I. M. Yaglom, B. A. Rozenfel'd and E. U. Yazinskaya, Projective Metrics (Russian), Uspehi Mat. Nauk 19 (1964), 51-113.

University of Missouri
Columbia Missouri

University of Southern California
Los Angeles, California

